THE ASTEROID HUNTER

THE ASTEROID HUNTER

A SCIENTIST'S JOURNEY TO THE
DAWN OF OUR SOLAR SYSTEM

DANTE S. LAURETTA

GRAND
CENTRAL

New York Boston

Grand Central Publishing
Hachette Book Group
1290 Avenue of the Americas, New York, NY 10104
grandcentralpublishing.com
twitter.com/grandcentralpub

First Edition: March 2024

Grand Central Publishing is a division of Hachette Book Group, Inc. The Grand Central Publishing name and logo is a trademark of Hachette Book Group, Inc.

The publisher is not responsible for websites (or their content) that are not owned by the publisher.

Grand Central Publishing books may be purchased in bulk for business, educational, or promotional use. For information, please contact your local bookseller or the Hachette Book Group Special Markets Department at special.markets@hbgusa.com.

Print book interior design by Marie Mundaca.

Library of Congress Cataloging-in-Publication Data
Names: Lauretta, D. S. (Dante S.), 1970- author.
Title: The asteroid hunter : a scientist's journey to the dawn of our solar system / Dante S. Lauretta.
Description: First edition. | New York, NY : Grand Central Publishing, 2024. | Includes bibliographical references and index.
Identifiers: LCCN 2023040980 | ISBN 9781538722947 (hardcover) | ISBN 9781538722961 (ebook)
Subjects: LCSH: Lauretta, D. S. (Dante S.), 1970- | OSIRIS-REx Asteroid Sample Return Mission. | Planetary scientists—United States—Biography. | Planetary science—Research. | Bennu (Asteroid) | LCGFT: Autobiographies.
Classification: LCC QB36.L32 L38 2024 | DDC 523.44—dc23/eng/20231012
LC record available at https://lccn.loc.gov/2023040980

ISBNs: 978-1-5387-2294-7 (hardcover), 978-1-5387-2296-1 (ebook)

Printed in the United States of America

LSC-C

Printing 1, 2024

To Kate, Xander, and Griffin. This journey belongs to all of us.

CONTENTS

CONTENTS

PROLOGUE

BENNU

THE MOST DANGEROUS ROCK IN our solar system is asteroid Bennu.

With a diameter about the height of the Empire State Building, Bennu is as massive as an aircraft carrier. Reflecting only a tiny fraction of the sunlight that shines upon its surface, it is also one of the darkest objects in our solar system. Most other asteroids reflect five times as much.

Bennu was discovered on September 11, 1999, by scientists at the Lincoln Laboratory at MIT, an entity tasked with keeping an eye on the sky, searching for potential threats from both foreign nations and the interstellar abyss. Bennu was of immediate interest because its dark surface suggested a carbon-rich composition, meaning it is a rare type of asteroid that could provide a wealth of information about the essential components of life and the foundations for a habitable world. Billions of years ago, others like it may have delivered the chemicals that make up the biomolecules in our cells, the water we drink, and the air we breathe.

Today, scientists are interested in Bennu because it poses a major threat. On September 24, 2182, if humanity takes no steps to prevent it and the odds line up, Bennu will hit the surface of the Earth at a velocity of Mach 36, or 27,000 miles per hour—a freight train crashing into the planet. Its path will blaze through the atmosphere many times brighter than the midday sun. The impact will release a blast of energy equivalent to 1,450 megatons of TNT. To put that in perspective, the total energy expended during all nuclear testing throughout history is estimated to be 510 megatons. Bennu's crash landing would triple that in an instant.

In some respects, the Earth would hardly register such an event: the orbit and axis would remain unperturbed. In other respects—arguably more pertinent ones—the consequences would be devastating.

Bennu's impact would create a crater four miles wide and half a mile deep. It would trigger a magnitude-6.7 earthquake. Approximately fifteen seconds after impact, areas within tens of miles of the crater would experience an air blast driven by Bennu's hypersonic path through the atmosphere and massive energy release at the surface. The wind would be twenty times more powerful than a Category 5 hurricane. The sound wave would be louder than an orchestra of air-raid sirens, blaring from every direction. Curious onlookers who flocked to their windows to view the fireball would be greeted with a barrage of fragments as the glass imploded. Residential homes would be flattened, the few survivors determined by location and random luck. Office buildings and highway bridges would twist, distort, and ultimately collapse. Trees would be blown down; those that somehow managed to remain upright would be stripped of their branches and leaves.

After another fifteen seconds—still well under a minute after Bennu's initial impact—fragments of earth and rock that had been violently excavated by Bennu would rain down upon this damaged area. The largest rocks that Bennu sent flying would be the size of sixteen-story buildings. In the aftermath, power outages, food and water shortages, and communication blackouts would last months as the region remained inaccessible.

In short, a Bennu impact would be a major natural and humanitarian disaster. Most of the damage would be concentrated within tens of miles of the impact site, but there would be catastrophic results for hundreds of miles. If the asteroid struck a major population center, the loss of life would be staggering.

Bennu's orbit brings it extremely close to our planet. It is this proximity that gives us a chance to determine whether we should prepare for disaster. If this space rock is going to hit, humanity will face a difficult choice: begin planning to evacuate a wide region of the world or launch a mission to knock the asteroid from its path. Either way, we'll need to know all we can about Bennu.

In 2011, NASA awarded me a billion dollars to accomplish just that. The mission would come to entail not only sending a spacecraft to the asteroid but bringing a piece of it back to Earth.

It is a story four and a half billion years in the making.

PART I

TWO CARBON ATOMS

LONG AGO, TWO CARBON ATOMS were born entangled in the heart of a star.

It was an aging star, nearing the end of its days. All its life, the star had been able to fend off the relentless pull of gravity. At first, this was easy. Grab any four hydrogen atoms and fuse them together until helium ash formed, generating enough pressure to negate gravity's tug—a brilliant steady state. Hydrogen was so abundant that the star felt like it would live forever.

But the hydrogen was not infinite, and one day, three billion years after it was born, the star ran out. Gravity is patient; it always wins. Slowly, it yanked the outer layers of the star inward. Eventually, the pressure in the star's heart became so great that the helium began to burn, producing new elements: nitrogen, oxygen—and carbon, including those two identical, entangled atoms.

The remaining helium burned more quickly than its hydrogen predecessor. After one billion years, the star was drained. It did not go quietly. Its cache of remaining hydrogen and helium reached nuclear temperatures in random pockets. The resulting surges of energy flung star fragments in all directions, releasing the carbon twins into space.

The two atoms sped away from their shattered parent star. They cooled and crystallized. They met up with other carbon atoms and banded together, forming chunks of graphite. The graphite wandered through the interstellar medium for millions of years, until it was swept up into the gravity of a new star, with a pull so powerful it commanded a swirling halo of gas and dust.

The graphite combusted with oxygen in the primordial disk and two new molecules of carbon dioxide sped away into space. The twins yearned to stay together but the swirling disk of gas and dust was turbulent and unyielding. The carbon atoms could not hold on to each other. Torn apart, they each collided with metal grains scattered in the dust clouds surrounding them. The metal surface catalyzed a chemical reaction, transforming each twin into sticky tar. Their tar-coated metal stuck to nearby dust grains, allowing them to grow into pebbles. The pebbles grew to boulders and the boulders grew to asteroids.

The carbon twins traveled inside their two host asteroids as they tore through space in random directions. One asteroid was flung inward toward the Sun. Finding other bodies like itself, it snowballed into a massive planet. This terrestrial carbon twin settled into the crust of the newborn Earth. Billions of years later, it became folded into the genetic code of a human being named Dante Lauretta.

The wandering twin avoided this fate. Its host asteroid survived the chaos of the early solar system. A billion miles from its sibling, it settled into a relatively stable orbit in a belt between Mars and Jupiter. The two carbon atoms were destined to meet again on September 24, 2023, when the OSIRIS-REx spacecraft delivered a sample of asteroid Bennu into Dante's waiting hands.

CHAPTER 1

ORIGINS

IN THE LATE SUMMER OF 1992, I was working as a short-order breakfast cook at a dive bar called Mike's Place in Tucson, Arizona. In addition to their infamous nickel pints and quarter pitchers of Old Milwaukee beer, Mike's offered a $0.99 breakfast—two eggs, hash browns, and toast—perfect for sopping up a night's worth of alcohol. Around 10:00 a.m., the University of Arizona frat boys stumbled in, like bears out of hibernation, delirious and hungry. They ordered three or four breakfast specials each. Behind the grill I flipped eggs like a machine, deftly dodging missiles of splattered oil.

Serving up breakfast had become second nature. I had been paying my way through college working jobs like this for the past four years. As my body moved through the motions of flipping hash browns and cracking eggs, my consciousness was free to ponder other things.

I was worried. I was about to start my fifth and final year as an undergrad. Only weeks before, left arm dangling from my beloved firetruck-red 1972 Volkswagen Microbus (nicknamed "Gus"), I had rolled back into Tucson after one final summer following the Grateful Dead, camping in the Sierra Nevada mountains around Lake Tahoe, and cooking out of my van for weary thru hikers on the Pacific Crest Trail.

My fifth year. That thought resonated in my mind. I never planned on spending five years in Tucson; it just sort of worked out that way. At the end of the last school year, I wasn't ready to move on. I had completed all the requirements for a theoretical math major, but I knew that wasn't my path. By taking a fifth year of courses, I could explore physics and Japanese culture—and also buy a little more time to save money and put off those upcoming student loan payments for another year. I was hoping to find clarity before embarking on my next stage in life.

I had grown up way off the grid, in a single-wide trailer that stood on tripods at the end of a long dirt road. The front yard was dirt, the backyard was dirt, and the Arizona desert surrounded us like an untamed wilderness. We had to drive miles to fill our five-hundred-gallon tank of water every couple of weeks.

Family life wasn't great. My father embarked on a steady diet of alcohol and marijuana until my mother kicked him out when I was twelve. I became a fatherly older sibling to my two little brothers, who were six and eight years my junior.

Dante and a friend on tour with the Grateful Dead circa 1992.
Credit: Dante Lauretta personal photo

My escape came from exploring the alien world of the desert. I collected scraps of metal to build a fort for myself and my brothers, a safe hideout for when our parents were fighting. I scouted for swimming holes and hunted for gems and minerals. In the desert, I sought wonder and distraction—and found plenty of both.

I especially loved picking through the remnants of old mining operations, imagining the men who had once staked out this desolate patch of land and set about excavating the unlikely cache, sleeping on the hard ground and rising with the sun. I would comb the tailings piles, feeling a thrill of discovery whenever an alluring mineral specimen emerged from the debris.

In the desert I could reinvent myself. I could pretend to be anyone. I was a swashbuckling adventurer, an explorer of the unknown. On one memorable expedition, I convinced one of the few neighbor kids to follow me into an abandoned mine. My friend was nervous, but I plowed ahead, displaying the kind of bravado I only wished I could show at home. Moments later, I heard coming from behind me the unmistakable rattle of a snake that meant business. I waited in the tunnel for hours until my friend returned with a stranger from the nearest homestead who quickly dispatched the snake with his pistol. Far from considering the afternoon a failure, I reveled in the memory of my most daring adventure yet.

I'm not certain whether the chaos of my young life necessitated this persona, or if it simply revealed it—but however it happened, it stuck. Even after my father left, even after my mom moved us to a Phoenix suburb and life evened out a little bit, I still wanted to test the limits of everything, to see how far into the wild I could go.

Like most smart kids in tough situations, I intuited that higher education was a way out, a way off the treadmill of endless financial struggle and unaffordable desires. Since no one in my family had been to college, it had taken on an almost mythical status in my mind.

Now, after four years of study, I found myself at a crossroads. Soon, I would graduate from the University of Arizona, and I had no idea what to do after that. I was looking for a spark, something that would ignite my curiosity and challenge my mind…and pay the bills.

As the breakfast crowd thinned out, my mind snapped back to the kitchen. It was time to clean up and head home, and hopefully get some of this grease out of my hair.

It was a stunningly bright Saturday in September when I untied my apron and stepped outside the diner. Nine hours in front of the grill was hell on the body, and I sat on the patio to rest my feet before walking home, a nickel pint of Old Milwaukee in one hand and Friday's edition of the *Arizona Daily Wildcat,* the student newspaper, in the other. I took a long drink and opened the paper wide. That's when I saw a full-page ad in blocky, decisive letters:

WORK FOR NASA

I pushed the sweaty hair out of my eyes to read the fine print, amazed that such a possibility even existed.

Those three words echoed in my mind. NASA represented the best of the best—the one part, I felt, of the US government that did very difficult things and did them very well. They asked big questions and launched big, bold spacecraft to seek answers. To work for NASA meant to be at the spearhead of human exploration.

It was as if the dirty window I had been looking through my whole life was wiped clean. I had found my path.

I headed home and immediately started working on my application to the NASA undergraduate research program, hoping beyond hope to land that job. Not only did I get it, but it changed my life's whole direction.

✦ ✦ ✦

Thirty-two years earlier, on a cool April night in 1960, a twenty-nine-year-old scientist in browline glasses named Frank Drake went to work at the National Radio Astronomy Observatory in Green Bank, West Virginia. He pointed an eighty-five-foot dish toward Tau Ceti, a star about the size of our Sun, roughly twelve light-years away. Dubbing his experiment Project Ozma, in honor of the fictional Princess of Oz, he tuned the receiver to 1420 MHz, the frequency of radio waves emitted by interstellar hydrogen. This frequency is in a part of the radio spectrum referred to as the "water hole," a cosmic quiet zone where both hydrogen and hydroxyl (one hydrogen and one oxygen atom) are detected. The "H" and "OH" lines, as scientists refer to them, are visible from anywhere in the cosmos, in the quietest part of the entire electromagnetic spectrum of the universe.

Drake was looking for signs of an extraterrestrial civilization. For the next two months, for six hours a day, he waited for a message from outer space. The search for extraterrestrial intelligence, or SETI, had begun.

Drake wasn't the first person to consider using radio waves for interplanetary communication; some of the technology's most notable pioneers believed they had received messages from extraterrestrials. While experimenting alone in his Colorado Springs lab in 1899, Nikola Tesla intercepted strange, unidentifiable signals. He became convinced he "had been the first to hear the greeting of one planet to another." In 1920, Italian inventor Guglielmo Marconi said he had received something similar. "I believe it is entirely possible that these signals may have been sent by the inhabitants of other planets to the inhabitants of Earth," he boldly stated.

In 1959, Giuseppe Cocconi and Philip Morrison published a seminal paper in *Nature*, in which they proposed scanning a narrowband frequency across the galaxy to search for life. Just as we had let our own radio and television broadcasts drift into space for years, Cocconi and Morrison reasoned that any intelligent civilization would be doing the same.

While Drake's two-month search was a bust, it provided a much-needed spark to the scientific community. A year after his project ended, Drake invited a group of ten astronomers to West Virginia for the first significant conference on extraterrestrial intelligence. Morrison was there; Carl Sagan was too. So was a man named John Lilly, a physician and philosopher working on communicating with another form of alien intelligence—dolphins. Understanding how animals talk to each other, he reasoned, was a step toward figuring out how creatures on other planets might as well. So taken were the attendees with Lilly's work that they decided to call themselves the Order of the Dolphin. After their time together, Lilly sent them all small silver pins featuring the marine animal.

Playful names aside, the conference's lasting contribution to SETI studies was the Drake equation, a formula that estimates the number of potential intelligent civilizations in the galaxy:

$$N = R^* \bullet f_p \bullet n_e \bullet f_l \bullet f_i \bullet f_c \bullet L$$

N: The number of civilizations in the Milky Way galaxy whose radio emissions are detectable.

R*: The annual rate of formation of stars suitable for the development of intelligent life.

f_p: The fraction of those stars with planetary systems.

n_e: The number of planets, per solar system, with an environment suitable for life.

f_l: The fraction of suitable planets on which life appears.

f_i: The fraction of life-bearing planets on which intelligent life emerges.

f_c: The fraction of civilizations that develop the technology to produce radio waves.

L: The average number of years such civilizations produce those signals.

The equation wasn't particularly practical to begin with; scientists didn't have concrete numbers to plug in for many of these variables. Instead, the Drake equation gave scientists a laundry list of problems to tackle. Refine the variables in this equation, and only then could we seriously consider how to find and communicate with other forms of intelligence, our cosmic neighbors in the Milky Way.

It took a decade and a half, but NASA finally put some skin in the SETI game in 1975, when it began funding projects around the country. Mercilessly disparaged in Congress as a farcical waste of taxpayer dollars, NASA's SETI initiatives limped along, underfunded and underappreciated. Eventually, nearly twenty years later, the science community got clever. Instead of asking for cash under a SETI heading, they tweaked the program ever so slightly and came up with a respectable new name. On October 12, 1992, a couple weeks after I saw that ad and rushed home to apply for the Space Grant program, NASA officially launched the High Resolution Microwave Survey.

As soon as I knew of its existence, I was a SETI fan. The concept of scanning the skies for aliens sounded like a high-tech version of what I had been doing for years. Sitting around a campfire, chemically modified, grappling with those giant questions: Why is matter alive? Why is it arranged in such a way that a pile of it is sitting here even asking this question? How could this phenomenon be limited to just this tiny speck of dust in one tiny corner of the universe?

My roommate Zac and I were completely aligned on this front. Even though we were cash-starved students, we pooled our money to buy a subscription to *Omni* magazine, a publication that delightfully blended science and science fiction. It was candy for a young scientist's mind. The October 1992 issue contained an article by the guru himself, Frank Drake, describing the NASA-funded SETI program. The opening paragraph was pure scientific swagger. In it, Drake coolly predicted that scientists "will probably discover intelligent life on other planets by the year 2000." As I read the words aloud to Zac, our eyes grew wide.

Later that month, I finally received news on my application in the form of a letter from a place called the Lunar and Planetary Laboratory. Puzzled, I told Zac to put his book down. "You ever heard of this?" I asked, holding the letter out so he could read the return address.

"That must be the Department of Planetary Sciences. I think I've seen their courses listed in the catalog."

"How have I never heard of that before?" I thought aloud as I hefted the volume from the bottom of the bookshelf and flipped to the *P*s. Sure enough, crammed onto a half page of the catalog was a stunning list of courses: Exploration of the Solar System, Jovian Planets and Satellites, and Principles of Cosmochemistry (whatever the hell that was). Somehow, I'd been a student at UA for four years, taking courses in physics, math, astronomy, and geology, and had no idea this department existed.

Remembering the envelope balanced on the arm of the couch, I chucked the catalog back onto the shelf. Hands twitchy, I tore it open. "It's about my application!"

"Congratulations!" the first line began. "You have been

selected for the Arizona NASA Undergraduate Research Space Grant Program."

"Holy shit." I passed the letter over to Zac as I collapsed on the couch. "I did it. I actually got the job." My grease-soaked days as a breakfast cook were over. I tried to imagine what kind of project I might work on as my eyes settled on the cover of *Omni*, its corners creased and well-worn. A bold block-letter font promoted the cover story, "On the Origin of the Solar System," by Carl Sagan and Ann Druyan, while a futuristic-looking spacecraft burst forth from interstellar flames.

✦ ✦ ✦

Dr. Carl DeVito's office was stereotypical of an academic, cluttered with stacks of papers, shelves packed with mathematical journals, the air suffused with the smell of old books and copier ink. I thought I knew the math department well; it had been my home at UA. I tutored in the math lab and used their computers for my homework. Yet I had never met Dr. DeVito, who was now the advisor for my Space Grant project, even though he had been a professor there for years. DeVito had been working with Richard Oehrle, a professor in the linguistics department. Together, they developed a logic-based language to communicate with alien societies.

As I settled in, Dr. DeVito radiated enthusiasm from behind his desk.

"It's an exciting time for SETI," he said.

I smiled and nodded, trying not to let my face spread into too wide of a grin—I didn't want to look like a noob.

But the fact that I was talking with an esteemed professor about communicating with aliens—and getting paid for it—had me squirming in my seat.

"When we do contact extraterrestrial life, we'll need a way to communicate with them. That's why you're here," Dr. DeVito announced.

I knew this much. My application was successful mainly because of my unique combination of majors. I had been studying science alongside language for years. DeVito's passion project required the ability to think creatively in both disciplines.

As he defined the project's scope, I could sense that we thought alike. DeVito treated SETI communication like a math problem. And, like any good math problem, he began with the assumptions.

"Our current SETI efforts are focused on searching the water hole," he started.

I nodded. This had been in Drake's article in *Omni* that I had nearly memorized.

"As a result," DeVito continued, "any contact between our civilization and an alien one would be via radio in this frequency range. This assumption immediately implies that our correspondents have a basic knowledge of science—enough, at least, to build a radio transmitter. By that logic, they *should* be able to learn a language based on fundamental scientific principles."

Any alien capable of building a radio transmitter must know all about the propagation of light waves through space. Such devices were complicated pieces of equipment, requiring detailed knowledge of electricity and magnetism. They had to understand the same fundamental facts about the universe that allowed earthly scientists to build their radio telescopes.

"If they can develop radio technology," DeVito continued, speaking faster as he realized that I was keeping up with his line of thought, "they can count, they know about chemistry, they are familiar with melting and boiling points, and they understand the nature of gases. If we share all this common knowledge, then we can communicate about numbers, chemical elements, and physical units such as the gram, the calorie, the degree, and so on. Once this basic set of concepts is established, more interesting information can be exchanged."

"How will you communicate information about the chemical elements?" I asked.

"The language is based on mathematical set theory," he replied. "The elements will be introduced as a new set, once the integers have been established. We'll start with hydrogen. Hydrogen is defined as the set of all atoms having atomic number one. Helium is the set having atomic number two, and so on."

"What if that's too simplistic?" I asked, surprised by my own boldness but wanting badly to impress my new mentor.

"I think we can do much better than that," I continued, "and maybe the answer lies in the SETI approach itself. The H and OH spectral lines that define the water hole show us how energy is partitioned in these two molecules. The lines make unique patterns and that is how we know that space is not an empty void—it is a chemistry laboratory.

"A better set," I offered, "is to use the hydrogen energy levels. Instead of assigning each element a number, which could be difficult to interpret, we could use their spectral 'fingerprint.' Any alien civilization that is exploring the universe is going to have their share of spectroscopists, and they are going to recognize these patterns immediately. Heck, they likely have libraries of these patterns to analyze their radio data."

DeVito nodded. "Sounds like we have a focus for your project. Go advance the DeVito-Oehrle language to transmit the periodic table!" he said theatrically.

I spent the rest of the school year savoring every moment that I worked on the SETI project. DeVito was an encouraging, invested mentor, and we met often at the Fidlee Fig, the breakfast buffet in the student union. Over the course of the project, I built what I called the "SETI Word Processor," a program that allowed the user to type in any chemical reaction and have it translated first into the DeVito-Oehrle language and then into binary. I imagined that someday my code would be attached to radio telescopes around the world, with teams of scientists typing out messages to our cosmic neighbors. In my wildest dreams, I was the technician pushing the button, sending a message to the stars.

Just a few months after I started working on this project, it was clear I would never leave. I decided to go to graduate school to study planetary science with a focus on SETI. If I could get into UA, which had one of the world's best programs, then I wouldn't even have to move out of my apartment.

The University of Arizona is more than my alma mater. It is—as far as I and many others in planetary science are concerned—one of the spiritual homes of the discipline.

In 1960, the same year that Frank Drake launched Project Ozma, a Dutch astronomer named Gerard Kuiper founded the Lunar and Planetary Laboratory in a remote corner of the top floor of UA's Atmospheric Sciences Building. At the time, the study of the planets and their satellites had fallen out of favor; the rise of space photography in the early twentieth century inspired scientists to see and study—for the first time—heavenly bodies beyond our galaxy. While our understanding of extragalactic space grew by leaps and bounds, knowledge of the planets we share the solar system with faltered. Kuiper, however, spent his early career researching planetary science, as unfashionable as it may have been. He discovered that Titan, the largest of Saturn's moons, had an atmosphere of methane gas, revealed never-before-seen satellites of Neptune and Uranus, and detected carbon dioxide on Mars.

In the 1950s, Kuiper focused his attention on our nearest neighbor. For years, his fellow scientists—when they paid attention to the Moon at all—couldn't seem to agree on anything about it. Was it cratered because of ancient volcanic activity or because it had been pummeled with asteroids? Was the surface soft and fluffy or hard and crunchy? Nobody knew. Figuring that the history and physical properties of the Moon could tell us a lot about the formation and composition of our own planet, Kuiper began work on a lunar atlas project, gathering and taking new photographs to give scientists a baseline for studying our fascinating natural satellite.

Kuiper's timing was impeccable. After the launch of Sputnik in 1957, Americans once again turned their attention to our own solar neighborhood. Four years later, when John F. Kennedy proclaimed that Americans would walk on the Moon before the decade was out, Kuiper was, quite literally, the man with a plan—the only man, in fact, with any plan. His mission to map the Moon was suddenly a national priority. When Kuiper set up shop on UA's campus, he brought with him 14,000 pounds of books, paper, and instruments. Over the next decade, Kuiper and the scientists he recruited to work at the lab would play an integral role in the Apollo program. (It would take years for Kuiper to get the respect he deserved, though; long after it was founded, LPL was still derided as the "Loony Lab.")

But perhaps Kuiper's greatest legacy was transforming modern planetary science into an interdisciplinary pursuit. Alongside traditional astronomers, he filled the halls of LPL with physicists, geologists, and atmospheric scientists. This trailblazing streak at LPL would continue over the next half century, as the lab became a training ground for young planetary scientists and ultimately made UA the first university to control a spacecraft mission from its campus.

In 1992, at the end of the fall semester, with graduate school application deadlines looming, I made an appointment with the head of LPL, Dr. Eugene Levy. Levy was a daunting figure, a man with a tall forehead, cleft chin, and a reputation for not mincing words.

"I want to go to grad school here at LPL," I told him. "And I want to search for extraterrestrial intelligence."

Levy's response was swift and simple: "Hell no." Like a cosmic pinprick, his words deflated my dream. "That's career suicide. We probably won't even admit you to the program if you put that in your personal statement."

"What do you mean? SETI is a legitimate, NASA-funded project," I countered.

"Not for long," he said, almost chuckling. "Senator Bryan has pretty much declared war on the program."

Indeed, as soon as the NASA-led search got underway, the Senator from Nevada made it his mission to cancel its funding. "This hopefully will be the end of Martian hunting season at the taxpayer's expense," he gloated in a press release.

Instead, Levy encouraged me to study Mars. In the early 1990s, like the Moon before it, Mars was on the rebound after nearly two decades of neglect.

Launched in 1975, the NASA flagship Viking Project provided astonishing data about the red planet. The Viking armada consisted of two orbiters, each with their own lander, to scour the Martian surface and atmosphere and search for evidence of life on the planet.

I remembered July 4, 1976, very well. I was five years old, and the front page of the *Arizona Republic* had a picture of a NASA robot that had just landed on Mars. That was when I first felt the spark, the same one I experienced when I opened that student newspaper.

In the years since, the orbiters had revealed volcanoes, giant canyons, cratered terrain in the south, and smooth young landscapes in the north. The images also contained abundant evidence of ancient fluvial systems, with riverlike channel networks, signs of catastrophic floods, and landforms reminiscent of shorelines. The landers, however, failed in their ultimate goal; the life detection experiments proved inconclusive.

After years of observation and experimentation, scientists concluded that solar ultraviolet radiation saturates the surface of Mars. This damaging radiation, combined with the extreme dryness of the soil and reactive chemicals, made the Martian surface uninhabitable to any kind of life we're aware of.

This conclusion took the wind out of the sails of Mars exploration. Throughout the '80s, NASA focused on building the space shuttle, de-emphasizing studies of the Moon, Mars, and more distant solar system destinations. But the scientific community continued to press for Mars, arguing it contained a wealth of untapped and valuable information. A strong case was made for comparative planetology, built around the evidence of ancient water flows. Clearly, Mars's climate had been different in the past. If we were at all concerned about the future of planet Earth, we needed to understand how Mars could transform from a warm, wet paradise into the barren desert world reminiscent of George Lucas's Tatooine or Frank Herbert's Arrakis.

Eventually, NASA listened. As I sat in Tucson discussing my career options with Levy, Mars Observer was hurtling through space, hailed as NASA's triumphant return to the red planet.

Hesitantly, I agreed to consider Levy's advice. I applied to several schools around the country that had faculty members engaged in the Mars Observer mission.

A few months later, in the spring of 1993, I opened a thick envelope from Washington University in St. Louis, brimming with good news. I had been accepted to the Department of Earth and Planetary Sciences, awarded a graduate fellowship in the McDonnell Center for the Space Sciences—alleviating my concerns about money for a while—and earned a research position with Professor Ray Arvidson, lead data scientist on the Mars Observer mission. Though I was also accepted to UA, Levy encouraged me to go to St. Louis and broaden my perspective.

So, I planned my move to the Midwest. I sold Gus, packed everything I owned into two tiny suitcases, and booked a plane ticket to St. Louis. My own personal transformation from rambling explorer to serious scholar had begun.

Before I left Arizona, I went on one last hike to say farewell to the desert. That evening, I sat alone on a summit in the Tucson Mountains, listening to wind whisper through the spines of the saguaro cacti around me. My eyes teared up as I thought of leaving this place, the only home I'd ever known and the place where I felt the safest. I lay back on the boulder and stared up into the dark sky. The stars winked back at me brilliantly and I could feel their summon.

CHAPTER 2

SIGNS OF LIFE

STANDING ON THE FLOODED BANKS of the Mississippi River, which, to this desert boy, seemed miraculously full of water, St. Louis had the aura of a *Big City*. Without mountains on the horizon to orient me, I felt lost in the decaying patchwork of old factories, confounded by the blatant racial segregation and Sunday blue laws.

Still, there was opportunity ahead of me at Washington University and I was grateful to be there. I strolled the Danforth Campus, which housed most of the academic departments, including Earth and Planetary Sciences, taking in the buildings' Collegiate Gothic architecture. I spent many afternoons admiring its square towers, corner turrets, and arched passageways. I found myself staring back at the gargoyles perched on corners, above doorways, and along windowsills. The contrast with UA couldn't have been starker.

Dr. Ray Arvidson had lured me to St. Louis with the promise of being at the forefront of modern Mars exploration, supporting the first team to study Mars since the Viking 1 lander ceased communicating with Earth in 1982, the result of human error that caused the lander's antenna to go down.

The transition from undergraduate to graduate school was a revelation. In Tucson, I had indulged every interest, studying East Asian culture alongside hard-core physics classes. However, in St. Louis, my sole focus became the world of space exploration, poring over every new discovery, discussing planet formation over lunch, obsessing over this new wave of Mars exploration of which, remarkably, I was a part. Gone were the directionless days of flipping eggs; I was involved in serious research—and doing my best to become a serious researcher.

When I arrived in St. Louis to join the Mars Observer mission team, the spacecraft was scheduled to enter orbit around the red planet in just over a month. I could hardly contain my excitement and anticipation as I thought about what lay ahead. The spacecraft was equipped with an impressive array of instruments to study the composition of the Martian surface, map the topography and gravity, search for a magnetic field, observe the weather and dust storms, and explore the structure and circulation in the atmosphere—tools to study Mars's atmosphere, climate, geology, and the characteristics of its moon, Phobos. But more importantly to me and many others, it was a mission that held the promise of unlocking some of the greatest mysteries of the solar system.

At the Space Grant Symposium in Arizona the previous spring, I had listened to a keynote speech by Phil Christensen, the principal investigator for TES (short for Thermal Emission Spectrometer), one of the six instruments on board. TES measured the heat emanating from the Martian surface. Amazingly, these infrared photons encode information about the mineral content of surface rocks, frosts, and the composition of clouds. A mere postdoc, Phil in his presentation took us on the roller-coaster ride of building, testing, and delivering TES to the spacecraft. I sat in awe in the audience, realizing that by winning the contract to launch a science instrument into space, Phil had also launched his career.

During my time at LPL, I also had the opportunity to attend a seminar by Bill Boynton, a professor who also had an instrument on board Mars Observer. His gamma-ray spectrometer measured the high-energy photons produced by nuclear reactions on the Martian surface. With the instrument's neutron detectors, the plan was to map the distribution of hydrogen—and therefore water—across the entire surface. The Viking experiments were thought to have disproven the existence of life on Mars, but many of us in the field remained unconvinced. One of the experiments had positively identified organic compounds on both Viking landers, but they were later shown to be chlorine compounds, interpreted as contaminants from cleaning fluids. Still, we held out hope that Bill's instrument would reveal where to focus future searches for evidence of life on Mars.

These two scientific instruments aboard the spacecraft were designed and built in Arizona, my home state. I felt a surge of pride knowing that those instruments represented the ingenuity and skill of my fellow Arizonans. It was a personal connection to the mission, and it only fueled my passion and dedication to making Mars Observer a success.

As we prepared for orbit insertion, Ray exuded an air of enthusiastic anticipation, like a parent awaiting the birth of their child. A veteran of the Viking program, he had devoted decades of his life to the study of Mars. The Mars Observer mission had been funded by NASA in 1986, just four years after the last signal from Viking. Ray, Phil, Bill, and countless others had been working tirelessly ever since. At that time, NASA's Mars Exploration Program was in its infancy, and the agency was seeking to develop a series of missions to explore the planet. Mars Observer was meant to be the first, one that aimed to gather data and pave the way for future human missions to Mars. At a total cost of more than nine hundred million dollars, the future of NASA's planetary exploration program was on the line.

To be part of a team dedicated to expanding our knowledge of Mars and pushing the boundaries of space exploration was obviously a dream come true. As we prepared for orbit insertion, the enthusiasm surrounding the mission was palpable, in St. Louis and around the world. The Mars Observer was more than just a vehicle fueled with hydrazine; it carried with it the collective aspirations of humanity to explore and understand the universe beyond our planet.

Beyond the invigorating work, one of the earliest bright spots of my life in St. Louis was the social scene. The first time I saw Kate was during my visit the previous winter at a departmental party for prospective graduate students. She strutted in with the air of an established scholar, a pair of hockey skates over her shoulder, her cheeks glowing from the rink. Kate blew right past me without a second glance, taking her place with the other graduate students, and I had not been able to get her out of my mind since.

As classes began, the graduate students gathered daily for lunch around a giant mahogany table in the department's main conference room. For weeks, Kate was a no-show, and I started to wonder if the beauty with the skates had been an apparition. Finally, one day, Kate reappeared. Everyone sat rapt as she described being waylaid in the Adirondacks, where she had been collecting bedrock from remote mountaintops. The next day in my mineralogy class, Kate assumed her duties as our teaching assistant, and I suddenly became very invested in the subject.

Kate was every bit the explorer I was, a brilliant geologist with a hunger for unlocking the secrets hidden deep in the Earth. But she was also steady and centered, the product of an idyllic upbringing in rural Connecticut, the only girl among her parents' four high-achieving children. When I looked into Kate's blue-green eyes, I saw an improbable combination: a lifetime of adventure and the possibility of building the family I never had.

Even though I missed the desert, the proud saguaro, the sooty smell of the creosote, the dark, star-filled skies, I had found new love with Kate and new meaning with the team leading humanity's triumphant return to Mars. Life in St. Louis was turning out all right.

That enthusiasm was tempered when, a couple of weeks later, I walked into Wilson Hall, intending to spend some hours in the basement office I shared with three other students. With orbit insertion happening the next day, I expected to find my colleagues in an anticipatory, if not celebratory, mood. Instead, I came upon Laura, a fellow aspiring Martian, collapsed against the wall, sobbing.

"What's wrong?" I asked, squatting down beside her.

"We lost contact with the spacecraft on Saturday," she blubbered. "The mission is over. My career is *over*."

My chest tightened. "Are you sure it's gone? Maybe it's just a communication glitch."

Laura looked up at me, ready to stamp out my optimism. "We've sent new commands every twenty minutes all weekend long—nothing. At first, the ops team assumed it had drifted off course, and that we would regain contact at some point. But it's been silent for over two days. The spacecraft should have recovered from any anomaly by now."

Months later, an independent investigation board from the Naval Research Laboratory would announce their findings: A fuel tank in the spacecraft's propulsion system likely ruptured, leaking fuel during the long cruise to Mars. The leak probably spun the spacecraft, causing it to enter "safe mode," and preventing it from turning on its radio transmitter.

"If it makes you feel any better, my plans are ruined too," I sighed, hardly beginning to register what this would mean for me.

"You've been here, what? Two months?" Laura retorted, "Big deal."

She was right. My two months were nothing compared to her two years, and it was infinitesimal compared to Ray's decades. I couldn't help but think of Bill and Phil, who had poured years of their careers into developing their scientific instruments. Their work had been meant to gather vital information about the red planet, and I couldn't imagine how they felt, knowing their instruments were now floating, useless, in space.

The loss of the Mars Observer was a stark reminder of the risks and uncertainties of space exploration. It was a harsh lesson that I would carry with me for the rest of my career. After the setbacks of SETI and the Mars Observer mission, I remained committed to exploring the universe and uncovering its mysteries, and I found myself searching for a new intellectual home. I needed a place where I could continue to pursue my passion for space exploration without the risk of Congressional interference or spacecraft malfunctions. It was a challenging time, but I counted my blessings and looked for new opportunities.

Fortunately, I didn't have to look far. One of my professors at WashU had a laboratory-based project focused on early solar system chemistry and planet formation. It dealt with one of the key variables in the Drake equation—the fraction of stars with planetary systems. I saw it as a legitimate path to fulfilling my SETI ambitions, and I eagerly signed on.

I was now working on understanding how the building blocks of planets form, shedding light on the origins of our solar system and the conditions necessary for life. Our first and biggest clue about the origin of the solar system is its very structure: the orbits of all the planets lie in a single plane and swing through space in the same direction. This configuration is the result of the "solar nebula," a swirling disk of gas, dust, and ice in the interstellar medium. As the cloud collapsed, its dense center attracted surrounding materials and created a disk that revolved around a growing young star. This is a phenomenon that physics calls "conservation of angular momentum." A great illustration of this effect is a figure skater performing a spin. When that skater begins to rotate, their arms are outstretched. As they draw their arms in, they are conserving their angular momentum, which causes them to twirl faster.

The same thing happens during star formation. Stars and planets are born from giant clouds in the interstellar medium. These clouds have dense masses at their center and their gravity attracts surrounding space dust, gas, and ice. Just like the skater pulling in their arms, the cloud spins faster and faster as it collapses, creating a disk that revolves around a growing young star. This stage lasts for a hundred thousand years or so, a geologic blink of an eye. Most of the materials in the disk ultimately go into building the star. Inside this disk, using the few scraps that remain, chemistry and physics work together to build planets, moons, and, at least on Earth, life.

The collapse of an interstellar cloud causes the system to heat up. The dust and ice that swirled there for millions of years are vaporized in a geologic instant. As the nebula ages, the disk cools down again and the very first planetary building blocks materialize. Just as snow forms when water vapor crystallizes, planet formation starts with condensation. In this case, the "snowflakes" are made of rock, metal, and sulfur, weather at solar system scales. It was this process—the condensation of matter from gas to the building blocks of planets—that I set about to study. My ultimate goal was now to figure out how Earth became a habitable world and how life gained its foothold here. In doing so, I hoped to unlock the secrets of our solar system's origins and the potential for life beyond our planet.

As I built an experimental system to study sulfur chemistry in protoplanetary disks, I marveled at the intricacies of planet formation, the delicate balance of forces that govern the universe, and how miraculous and ancient our solar system truly is. In many ways, it was a more grounded pursuit than SETI or Mars exploration, but it was no less thrilling. It was the start of my journey to the dawn of the solar system, and as a series of discoveries shook the world of planetary science, the significance of my work became even more apparent.

In the spring of 1993, husband-and-wife team Gene and Carolyn Shoemaker were scanning the night skies from the Palomar Observatory outside San Diego. Gene was a pioneer in astrogeology who worked under Gerard Kuiper during the 1960s to map the Moon. Convinced that icy comets had delivered water and other ingredients for life to Earth—and concerned about what might happen to civilization should one strike again—he was spending the last decades of his life investigating impact craters and conducting a systematic search for hazardous celestial bodies.

But it was ultimately his wife, Carolyn, a homemaker turned amateur astronomer at age fifty-one, who first laid eyes on the comet that became known as Shoemaker–Levy 9. (According to her New York Times obituary, it was "the only moment in her life when Ms. Shoemaker drank champagne straight from the bottle.") While Carolyn had previously discovered thirty-two comets (a world record), this one was different. Already yanked apart by Jupiter's tidal forces, it was now a fiery string of pearls hurtling through space at 134,000 miles per hour on a collision course with the king of the solar system.

For the first time ever, humanity had a chance to see a comet impact in action—to study what they are made of, how they travel through the solar system, and what effect they can have on a planet's atmosphere and interior. Not only would Shoemaker–Levy 9's impact result in fireworks like no other, but it would also allow us to better understand Jupiter, the Earth, and the role comets may have played in our origins.

Over the course of eight days in July 1994, all eyes, and all space instruments—including the Galileo, Voyager 2, Ulysses, and Hubble spacecrafts—turned to Jupiter as the comet's twenty-one fragments smashed into the planet with the force of three hundred million atomic bombs, earning the cover of *Time* magazine in what it called a "cosmic crash."

As I stared at the shocking display, the enormity of the event struck me. Jupiter wore proud scars where each splinter pummeled its surface. In infrared images, the impact glowed like a giant wart on Jupiter's lower left side. The wound continued to radiate as the planet rotated. As I marveled at the deep and lasting effects that Shoemaker–Levy 9 left in Jupiter's atmosphere, I couldn't help but wonder: *What would happen if the comet was coming for us instead?* I had always looked up at the night sky in awe. After Shoemaker–Levy 9, I also started to worry about our security and what humanity could do to deal with such cataclysmic threats from space.

The impact of Shoemaker–Levy 9 into Jupiter provided a new perspective on the importance of studying asteroids and their potential impact on the future of planet Earth. After all, if an asteroid hadn't come along sixty-five million years ago, then the mammals wouldn't have taken over the planet. Those thundering reptiles ruled the world for 140 million years before the cosmic hammer came down. Without that strike, there was no reason they wouldn't have continued to dominate the Earth. And if our ancestral mammals hadn't survived that Armageddon, then humans would not have evolved, radio telescopes would never have been built, and Drake's variable representing the fraction of civilizations that develop the technology to produce radio waves would be smaller by one.

While many astronomers focused on searching for potentially hazardous asteroids and comets, others dedicated their efforts to combing the skies for evidence of cosmic genesis. In October 1995, a pair of Swiss astronomers discovered the first exoplanet, or planet outside our solar system, located about fifty light-years from Earth. When Dr. Levy warned me away from my SETI ambitions, scientists knew of nine planets—the same ones we all memorized in grade school. The estimates of the fraction of stars with planetary systems went as low as one in ten thousand. With this revelation, I felt like the universe was daring me to explore it, just like that old mine shaft in the Arizona desert. Today, at least half of all nearby stars are known to host planets, a smorgasbord of possibilities for life.

As if everything was coalescing at once to set my path, in August 1996, a team of NASA scientists made a bombshell announcement that rocked the scientific community and the world at large. They had reportedly found evidence of primitive microbial life in a fragment of a

Martian meteorite that had been collected in Antarctica twelve years prior. It provided some of the strongest evidence yet that life could exist beyond Earth. The announcement made headlines around the world and even prompted President Bill Clinton to make a formal televised statement to mark the event.

The discovery was particularly meaningful for me because it highlighted the interconnectedness of planet formation and the potential for life to arise beyond Earth. The fossilized microbial life in the Martian meteorite also raised the possibility that the Viking experiments might have indeed detected life after all, intensifying my interest in, and giving a new urgency to, the study of the conditions that give rise to habitable planets. Any hesitation I felt in pursuing the search for life in the universe evaporated.

For the first time, science had produced something close to proof that life could exist outside of Earth, while in the midst of discovering vast new frontiers where we might go look for it. The excitement and buzz surrounding the discovery were palpable, and it launched the new field of astrobiology, which seeks to understand the origin, evolution, and distribution of life in the universe. The possibility of extraterrestrial life, even in the form of fossilized microbes, changed the way I think about our place in the universe. Maybe, I thought to myself, we are not alone.

I redoubled my efforts in the lab and the experimental work proceeded rapidly. After less than four years, I was putting the final touches on my dissertation. In the period leading up to that career-defining moment, there were many challenges that tested my resilience and determination. However, amid all the hard work and academic pressures, there was a bright spot of joy and happiness: the girl with the ice skates.

A few months before my defense, I decided to take a leap of faith and ask her to marry me. It was a bold move, and one that I wasn't sure I was ready for. But my heart told me that this was the right thing to do, and that she was the right person for me.

To my immense joy, she said yes. I remember the sensation of a weight being lifted off my shoulders. Suddenly, anything seemed possible. Looking back on that moment, I realize that it was a turning point in my life. It marked the beginning of a new chapter, one filled with love and hope. And even though the road ahead would be filled with trials and setbacks, I knew that I would never have to face them alone.

In 1997, a newly minted PhD—freshly engaged and as pale as I ever hope to be—I made my triumphant return to the Southwest, to start a postdoctoral research appointment at Arizona State University. This was a chance to sharpen my specialties, learn some new tricks, and publish some solid science.

Much like our solar system, Phoenix sits in the center of an orbit crowded by satellite cities. Tempe, where ASU is located, is one of the most beautiful. While many parts of Phoenix are parched and brown, Tempe uses flood irrigation to make the town miraculously lush and manicured, an oasis in the desert. The school itself is handsome and modern, with rows of palm trees towering over futuristic buildings. All this tidiness belied the fact that ASU was a wild place, a campus known for revelry. Students took great pride in earning the school a top spot on the annual "best party school" lists, and I was surprised to find the grown-ups didn't shy away from the debauchery. The faculty bashes I attended were not wine-and-hors d'oeuvre affairs; they were sweaty ragers, people packed into living rooms, dancing their asses off.

Two years after the major fanfare surrounding life in a Martian meteorite, NASA doubled down on their investment, dedicating nine million dollars for the inaugural NASA Astrobiology Institute, an interdisciplinary research effort made up of experts from across the country. ASU was one of the winning institutes, and suddenly, I found myself in a hotbed of astrobiology. My research trajectory had collided with the growing network of scientists dedicated to searching for life in the universe.

With the goal of securing a faculty appointment, I began seeking teaching opportunities to expand my skill set. Inspired by our large astrobiology grant, I joined a group of colleagues to offer a course called Cosmic and Biological Origins. In my section, I delved into the fascinating subject of planet formation, while other professors explored cosmology, the origin of life, solar system exploration, evolution, and the rise of human civilization. It was an exciting blend of disciplines, and as I sat through the other lectures, I began to see the connections between fields such as particle physics, chemistry, and biology. It broadened my horizons and solidified my interest in interdisciplinary research, where the intersection of different fields can lead to groundbreaking discoveries.

When I arrived in Tempe, I was fresh from a laboratory where I had spent years attempting to recreate the conditions at the dawn of the solar system to figure out how it came to be. My goal had been to produce a 4.5-billion-year-old weather report, and I was pretty sure I had done just that by fiddling with different gases and pressures and temperatures to grow primordial snowflakes of iron and sulfur. Now it was time to work backward, to study the rocks that were witness to the early solar system's violent transformations. For the first time in my career, I needed to get my hands dirty. I needed to score some meteorites.

Meteorites are rocks that have fallen to Earth from space, and they hold a wealth of information about the history and formation of our solar system. Fortunately for me, ASU is home to one of the world's greatest collections. In 1961, Dr. H. H. Nininger, a self-taught collector and expert, sold his extensive collection to the university, establishing the Center for Meteorite Studies. Inside the Center, there was a small, somewhat pathetic public display cabinet, as well as a conference room, decorated in classic academic gauche dark wood and red velvet accents. Past that, you'd find the treasure trove: the legendary vault where hundreds of rare and precious denizens from deep space resided.

Dr. Carleton Moore was the director of the Center, a rotund man with silver hair and a friendly face. Carleton was as welcoming and eager as an army recruiter, always ready to bring new researchers into the fold. On my first visit, I told him I wanted to see some primitive chondrites—the solar system's first sedimentary rocks. These objects formed by sweeping up dust floating through our protoplanetary disk and had remained essentially unaltered since. These were some of the rarest types, giving them high academic and economic value.

"Follow me," Carleton said, grabbing a ring of brass keys—the kind stamped with the words Do Not Duplicate—and charged down the hallway. My heart rate ticked up as I realized where he was leading me—the vault!

After Carleton unlocked and peeled open two sturdy doors, we stepped into the cool, rarefied air of the treasury. Metal shelving units lined the walls, with meteorites scattered across them like an interstellar art gallery. Along the far wall was a bank of wooden cabinets with a low table at the center, also adorned with giant samples. Everywhere I looked, rocks from space stared back at me, some in desiccators, some sliced in half to reveal their arresting interiors, others big, chunky specimens.

Carleton stood in satisfied silence as I took it all in, my eyes wandering over each surface, each sample. I spotted the well-known Canyon Diablo irons, fragments of the asteroid that smashed into central Arizona fifty thousand years ago, forming the mile-wide and world-famous Meteor Crater. I was drawn to these hunks of metal, minerals that crystallized deep in the molten metallic iron core of a long-gone protoplanet. Their surfaces were smooth and pocked with divots that looked like thumbprints, "regmaglypts" that formed as the iron hurtled through the atmosphere at supersonic speeds, ablating material as white-hot plasma from the heat of entry. I ran my hands over the samples, tracing out the contours, intuiting somehow that this iron wanted to be touched. (They do: The oil from our skin protects them, creating a barrier against the moisture that is always attacking and corroding them.)

Carleton led me over to a drawer labeled "ordinary chondrites." As he pulled open the ancient wooden drawer, I was greeted by a metallic, sulfur-laden scent that called to mind the emergency fire sticks I stashed in my backpack on camping trips. I didn't know if I was more amazed that meteorites have a smell, or that I was the one smelling them. I breathed deeply, pausing for a second to geek out about the molecules entering my body and the impossible span of time they had existed before becoming a part of me. We are all made of star stuff, as Carl Sagan famously said, and now I was inhaling just a little more, a little extra fuel for my scientific studies of their origins.

Seeing that I was a tactile learner, Carleton handed me a pair of white cotton gloves and I moved around the room caressing the stones, elated with such unfettered access to the holiest of scientific shrines.

And then something truly crazy happened. Carleton handed me the keys.

"Log your samples and be sure both doors are locked up tight when you leave," he said with an impish grin.

To begin my detective work, I loaded up on some of the most famous carbonaceous chondrites—Orgueil, Murchison, Mighei, Cold Bokkeveld—feeling like a jewel thief plucking the Hope Diamond out of the Smithsonian. As I held the meteorites in my hands, I felt deeply that these were no ordinary rocks—they had traveled through space for billions of years before finally falling to Earth, and they held secrets about the earliest days of our solar system. With my scientific treasure in hand, I carefully logged the samples and set about analyzing them.

As I worked with the electron microscope, the crystal

structures of the minerals within the meteorite were unlike anything I had seen before, and their composition told a story of the conditions under which the rock had formed. But it wasn't just the scientific data that captivated me; it was the beauty of the meteorites themselves. The intricate patterns of minerals and the sparkling metallic grains seemed almost like works of art, and I found myself getting lost in their intricacies.

Every sample was unique, and every analysis revealed new insights into the dawn of the solar system. They were a reminder of how vast and mysterious our universe truly is, and how much there was left to discover.

CHAPTER 3

HARVESTING THE STARS

WHEN THE CALL CAME THAT UA wanted to interview me for a faculty position, I was flattered, but not at all hopeful. My interview was with the new director of LPL, Dr. Michael Drake. A bombastic, overbearing Brit, Mike had big glasses and a big ego. He was a giant in the world of planetary science. I had first crossed paths with him briefly four years earlier at a conference in Maui. Serving as the president of the Meteoritical Society, Mike gave a brilliant lecture linking a certain type of meteorite to the giant asteroid Vesta. I listened intently as he explained how his team had meticulously crafted their science case, providing the first—and at that point the only—firm connection between specimens on Earth and a specific rock in space.

The interview went smashingly well. Mike and I hit it off and spent the better part of the day discussing all of the possibilities ahead of us if I accepted the job offer.

I went back home and told Kate, "I think we're moving to Tucson."

Just one short year after starting at LPL, I was invited to be part of the 2002–2003 Antarctic Search for Meteorites (ANSMET) team. By joining this historic campaign, I was hoping to gain a deeper appreciation of our planet and, of course, retrieve some amazing new samples from outer space. The ultimate dream was to find a stone particularly rich in carbon and associated organic molecules, something that might unlock the grand mystery of the origin of life.

While meteorites fall randomly across the globe, most of them end up in the ocean. Others plummet into remote places where humans rarely or cannot travel. The precious few that land in populated or accessible areas are often camouflaged among our own geological detritus. About seventeen meteorites touch down on planet Earth each day; 99.999 percent are lost forever. It is enough to break a scientist's heart.

Even when they end up in an easy-to-get-to location, time is not on our side. As soon as a space rock takes up residence on our planet, it is subjected to the elements: water, wind, and air. Microbes quickly colonize the carbonaceous chondrites, likely feasting on the same compounds that our earliest ancestors ate four billion years ago. Of course, their satiation diminishes our ability to peer back over these eons and try to piece together that history.

During my time at ASU, I had collected meteorites from the Arizona desert. These rocks were intact mainly because they had fallen into a region with an arid climate onto swatches of land that were shedding sediment, allowing the sample to weather slowly and remain unburied. In general, deserts are great places for hunting space rocks. In the Sahara, for example, a burgeoning trade in these samples has brought an influx of cash to residents and local economies.

For similar reasons, another great place to hunt for meteorites is on the glaciers of Antarctica. The first Antarctic meteorite was discovered in 1912, during what we call the Heroic Age of Antarctic Exploration, when men like Ernest Shackleton and Robert Falcon Scott explored and mapped the continent, risking—and often losing—their lives in the process. As a kid, I devoured stories of Antarctic adventure, imagining myself following in their footsteps to reveal some new geologic treasure.

A prospector named Francis Howard Bickerton found what was later named the Adelie Land meteorite, a one-kilogram stony object, during the storied Australasian Antarctic Expedition (a journey remembered for the incredible endurance of Douglas Mawson, who watched two of his friends die and had to eat his dogs to survive). The discovery of the Adelie Land sample was humanity's first indication that Antarctica held the richest meteorite field on Earth.

Geography is largely to thank for this phenomenon. The Antarctic ice sheet is like a conveyor belt, collecting fallen stars over thousands of square miles as it slowly flows out toward the edge of the continent. The Transantarctic Mountains act as a barrier to the ice sheet, essentially stopping it—and the treasures within—in their tracks. Powerful katabatic winds then get to work eroding that ice, eventually revealing the space debris that has been entombed below for hundreds of thousands of years and bringing it to the surface, creating concentrations at the foot of the mountains.

At that point, spotting the meteorites is as simple as knowing where to look. On the blue ice near the base of these mountains, dark rocks stand out among this blank canvas like a raisin in a loaf of white bread, and any stone you find out there can only have fallen from the heavens.

Fifty-seven years after the discovery of the Adelie Land sample, a survey team of the Japanese Antarctic Research Expedition found a concentration of Antarctic meteorites. In 1969, members of the crew gathered stones in the southeastern Mountains of Yamato, quickly realizing that they were probably holding valuable scientific material. Upon their return to Japan, the nine rocks they collected were indeed identified as meteorites. Even more tantalizing, each was distinct, representing a unique asteroid carrying its own saga of solar system history. This discovery prompted annual systematic searches by both Japanese and American expeditions.

The US-based ANSMET was founded shortly after by William Cassidy, a geologist whose name now adorns a glacier, a mineral, and an asteroid. It has occurred every year since 1976. From a pool of hundreds of applicants, ANSMET selects a crew of eight to twelve researchers—everyone from high school teachers to museum professionals—for the ultimate Easter egg hunt: six weeks kicking around the southernmost continent, searching for and snatching up meteorites. The program has collected tens of thousands of samples, including chunks of Mars and the Moon that were ejected during asteroid strikes. ANSMET's motto is "messis sidera," or "to harvest the stars."

The specimens garnered by ANSMET go to NASA's Johnson Space Center, where they are examined, cataloged, and curated. Researchers at the Smithsonian receive a slice of each sample to provide detailed classification information. The results are then published in a biannual newsletter that is distributed to research facilities around the world. Scholars and scientists can request samples to aid in their own scientific investigations. It is like the public library of meteorites, filled with untold wonders waiting to be uncovered.

This had earned ANSMET the nickname "the poor person's space mission." As ANSMET said, these objects were "delivered to us free of charge." All we must do is go get them.

✦ ✦ ✦

There were twelve of us in the 2002–2003 cohort, and we represented a wide range of interests and institutions. Our leader was Nancy Chabot, a research scientist at Case Western Reserve. I knew Nancy from her time as Mike Drake's graduate student at LPL, and she felt like academic family.

Cady Coleman was a NASA astronaut who had been to space twice. Cady had a big smile and a commanding presence. She loved to talk about her young son and her life on a farm in New England with her famous glass-artist husband. Carl Allen, also from NASA, was the head curator at the Johnson Space Center. He was a lean figure with a quiet but friendly style. A unique addition to the expedition was Andy Caldwell, a high school teacher from Colorado who was taking part in the Teachers Experiencing Antarctica Program. Andy was a science geek and was already blogging to his students back in Colorado. Linda Welzenbach, a short, spunky redhead with a passion for music and photography, was a curator at the Smithsonian. Danny Glavin—a California boy I nicknamed "the kid" and quickly adopted as my honorary little brother—was a recent PhD working on a postdoc at the Max Planck Institute in Germany. Danny had been involved in testing the Martian meteorite fossil-life hypothesis and was particularly interested in the possibility of organic contamination from the ice.

It was a sunny day in mid-November when we first gathered together in the bustling international terminal of LAX. We hugged and shook hands, figuring out our mutual friends or where we may have crossed paths before. We were all so different—different ages, backgrounds, areas of study—but in that moment, I could tell that we all shared the same wide-eyed expression of disbelief. We were going to Antarctica!

It took us the better part of a day to get to Christchurch, New Zealand. As soon as we unloaded our bags and got a bite to eat, it was time to get to work. Nancy led our first official briefing on the expedition plan. As she laid out the maps in front of us, I felt that familiar buzz of excitement that came from planning a long trip in the wilderness. I was always compelled to go where "no one had gone before." This trip into the Antarctic deep field seemed as close to that as I could possibly get.

"This year will be a little different," Nancy explained. "Because ANSMET has received extra funding, we are sending two groups out onto the ice. Four members will be heading to the LaPaz Icefield to do some reconnaissance work, scoping out the area to see if it is worth sending future cohorts to. The rest of us will be performing systematic searches in two locations over six weeks."

"The first, Goodwin Nunataks, is a well-known ice field that has already yielded hundreds of meteorites. It was last systematically searched in the 1999–2000 season, giving up more than four hundred specimens, but only half of the exposed blue ice was scouted. The second, MacAlpine Hills, is where we will spend most of our time. It lies north and west of Goodwin Nunataks. This area hasn't been searched in nearly fifteen years. Getting to these two sites will require multiple traverses, and we will haul thousands of pounds of gear via snowmobile across a hundred miles of ice."

We would be passing through some stunning terrain. The Transantarctic Mountains were exposed across our entire route. With some of these rocks dating back to before the dinosaur-killing asteroid, I looked forward to some "roadside geology of Antarctica."

Nancy gave us fair warning that the timing of each traverse and the days we spent at the collection sites would be up to the weather, military logistics, and the workload required to finish the job. With the itinerary set, our next step was readying ourselves to face some of the harshest conditions on Earth.

The following day, we gathered at the National Science Foundation's processing center, where the gear was displayed like the merch table at a concert. Coveralls and balaclavas, furs, and fleeces lined a barren wall as if someone had dissected an Antarctic researcher and carefully pinned and labeled all the various parts of their anatomy. Each one of us was issued a set of these garments, donning the very same clothes that our ANSMET predecessors had kept warm in on their own journeys. After receiving and trying on all my gear, I felt a bit like a Russian doll: peel away my parka to reveal my snow pants, which cover my fleece layer, which bulges atop my thermal layer. The look was completed by white rubber boots that Bugs Bunny himself would be proud to don.

The flight from Christchurch to McMurdo Station, Antarctica, was an eight-hour ordeal that we endured in the claustrophobic belly of a "Kiwi" C-130, courtesy of the Royal New Zealand Air Force. We were strapped onto cargo nets suspended above our gear, our knees nearly touching the person's across from us. As we taxied for takeoff, one of the passengers joked that the army used C-130s for paratrooper training because jumping out of one was preferable to sitting in it.

As the engines roared to life, I looked around at my compatriots and wondered if they were feeling what I was: nervous anticipation, the way your stomach drops as the roller coaster creeps up that very first hill. The point where you start to think, *Did I really want to go on this ride?* The "hill," in this case, was the "boomerang point": the place over the Southern Ocean where the plane no longer has enough fuel to return to New Zealand. Four hours into the flight, there was no turning back; it was McMurdo Station or bust.

McMurdo was run by the National Science Foundation, the entity that was funding our field campaign. It was the main "community" in Antarctica, with about 1,000 residents during the summer and 250 who stuck around through the punishing winters, when the mercury rarely inched above -20°F. Most residents were operations personnel, the people who made it possible for scientists like us to come and explore this quiet, frozen world. They worked in the harbor, in the heliport, or in one of the three airfields (flights are constantly taking off and landing, full of eager, jet-lagged researchers). They serviced the station's fleet of vehicles, served food in the dining halls, and restocked the community's markets. It was a wild existence, one our crew marveled at daily during our stay in this little town at the bottom of the world.

To me, McMurdo Station looked like a military base or a mining camp, but on the ground, it felt very much like a college campus. Not only was it teeming with scientists working on every discipline you can name, but the whole place also crackled with a familial excitement, not unlike a freshman dorm, and parties were easy to find. (If you're ever in Antarctica and looking to throw down, I would direct you to the Kiwi cargo lounge, where guests were treated to Speight's beer served with Marmite and crackers.)

During our long days (literally—the sun never set during this time of the year) at McMurdo, we trained. Before we could head out into the "deep field," we had a lot to learn about how to survive in Antarctica. The first thing our mountaineers Jamie and John taught us was how to pack our sledges, the eight-foot-long sleds that would carry everything we needed. The sledges were remarkably primitive; they looked nearly identical to the ones men like Ernest Shackleton hauled, only ours would be towed by snowmobiles, not huskies. Seeing them loaded up with our tents, food boxes, safety equipment, and survey gear, I had a newfound appreciation for those who made this journey without the assistance of a gas-powered motor.

Despite all our new-fangled equipment, our team would face the same dangers that explorers did a century ago: freezing temperatures, vicious storms, and miles-deep crevasses. One afternoon, Jamie and John sat us down for a long, terrifying lecture about all the things that can—and have—gone wrong in the field. They told us about a team member who was severely burned when a stove exploded in his tent, and of another who fell nearly one hundred feet into a crevasse. The session reminded me of the slideshow of car-crash photos I was subjected to in driver's ed, and I walked out of the conference room shaken but resolved.

Our training continued with an overnight excursion twelve miles afield, for us to practice setting up camp and rehearse rescue protocols. We learned how to secure a rope without a harness, how to climb up it with a Prusik knot, and how to set up a pulley system to rescue a teammate. Jamie and John tested our knowledge in a final exam of sorts when they lowered Carl into a crevasse and challenged us to get him out. At first, we struggled. We couldn't get our makeshift pulleys to work. When Carl accidentally dropped his glove, I watched in horror as it descended with an eerie silence, never audibly or visibly striking the bottom. As the day ended, we finally got our act together and "rescued" Carl from peril.

A week after we arrived in McMurdo, we were fully trained, and our gear was packed up and ready for transport. Still, there were four more days until our flight to Beardmore Glacier, where we would camp before a short traverse to Goodwin Nunataks. We spent evenings in the Kiwi cargo lounge and afternoons exploring the area around the station, including Robert Falcon Scott's old hut, which still had tins of haddock, veal, and cabbage lining the shelves.

That lonely outpost brought the historic nature of our work into focus. Humans had always pushed the boundaries, never content to stay within the well-known borders on the map. I felt part of this desire, this drive to continue to expand our knowledge and use our intellects to figure out our past and plan for a bright future.

On the day we were scheduled to depart, Jamie woke us up with a loud knock on the door, telling us our flight had been cancelled due to bad weather. The next two days began the same way. Each morning, my hopes sank as the information filtered through my sleep-fogged mind. We did our best to keep busy, hiking up McMurdo's famed Observation Hill to take in the sights and pay our respects at the base of the nine-foot cross that memorializes Scott and his men, who died during a 1912 expedition. The names of his crew were inscribed on the cross, as well as the final line of "Ulysses" by Alfred Tennyson: "To strive, to seek, to find, and not to yield."

Finally, two weeks after we arrived, the weather, the flight schedules, and our fate all lined up. Our plane was ready to transport us out to the field. We would be delivered to Beardmore Glacier in two groups, and mine was first. John, Jamie, Danny, and I boarded the ski-equipped LC-130 operated by the New York Air National Guard and took our "seats" in front of an impressive pile of cargo. To support six weeks of field operations, we were hauling more than four hundred pounds of food per person; a dozen stoves; heavy down sleeping bags with pads; duffel bags full of cold-weather clothing; four tents; eight snowmobiles; sixteen sledges; thirty bottles of booze; and enough fuel to cook, melt ice to make drinking water, and heat our tents. It was quite a sight to see all our material needs for the next six weeks packaged up, ready to be delivered to the heart of meteorite country.

After a two-hour flight to the Transantarctic Mountains, the plane descended for a bumpy landing on the surface of the uneven glacier. I felt the touchdown in the pit of my stomach, which was churning with anxious excitement. The tail of the LC-130 dropped down, and the cargo was deployed. I stared in awe as pallet after pallet flew from the back of the plane and slid to rest on the ice. Once all cargo had been ejected, the plane stopped, and the four of us emerged into the barren, alien landscape.

The Air National Guard could not leave the glacier until we got one tent set up and one stove burning. Danny—who would be my tentmate for the next six weeks—and I got busy setting up our makeshift home and igniting two stoves. Once we did, John lifted two thumbs in the air to let the pilot know we were all set. Several minutes later, the engines fired, and the plane peeled up from the ice, banking toward McMurdo, where the remaining four members of our team were waiting.

With the aircraft out of sight, the engines faded into silence. The four of us stood there watching the plane recede into nothingness. Soon, the only sound was the gentle breeze pushing small ice crystals along the glacier. I registered that the four of us were the only living creatures within a two-hundred-mile radius. There were no plants, no animals, no fungi; not even bacteria could survive out here.

I took a deep breath and inhaled a lungful of crystal-clear, cold Antarctic air, sensing a purity beyond any previous experience. The silence was foreboding, the sensation of isolation almost overwhelming. That's when it truly dawned on me just what I had signed up for.

A few days later, our group of eight had established our first base camp at Goodwin Nunataks, erecting four raincoat-yellow tents in a semicircle, and we set out on our first quest. To search, we lined up our eight snowmobiles about fifty feet apart, like some sort of glacial cavalry. Slowly then, we accelerated, trawling across the ice like boats in a harbor, scanning the blue ice in our individual "lanes." When someone spotted a sample, they hopped off their snowmobile and performed a highly scientific dance to attract the attention of the other team members. It was considered a major faux pas to cross into another hunter's lane to collect a meteorite, and as a rule, we all got credit for a discovery.

Collecting the meteorites was thrilling, but also something we took very seriously. We carried three collecting kits among us so that one was always nearby. The specimen was first given a number with an aluminum tag and photographed in place with that number on a manual counter. The dimensions and appearance were recorded, and it was then moved using a pair of sterile tongs. We did our best not to touch it or let it touch anything but the tongs. Still, accidents did occur, and we made a note of any deviation in the record book.

After the meteorite had been examined and documented, we placed it into a sterile plastic bag and sealed it with freezer tape. We brought the haul back to camp and put them in a storage unit called an isopod, which would be shipped back on the icebreaker ship that was due in port at McMurdo in January, when the Ross Ice Shelf receded enough for passage. We would repeat this procedure eight hours a day for six weeks, for as many days as the weather allowed.

Slowly but steadily, every other member spotted a meteorite and we all gathered to collect the sample. After several days, I had not spotted a single stone. I felt dismayed and a little concerned that I was somehow missing them. Was it possible that there was something wrong with my eyesight? Had I left some scientific treasure in my snowmobile tracks?

Finally, during a random foot search in a moraine, I spotted it: a nice little black rock that really stood out. As soon as I recognized the space stone, a sense of relief flooded through me. It was as obvious as the nose on my face.

After a week or so, I accepted that searching for space rocks was a lot like fishing: long, boring stretches interrupted by adrenalin-pumping excitement. The difference, of course, was that our "fish" could alter our understanding of the solar system and the secrets it may hold.

Thousands of specimens that teams like ours had found in Antarctica had proven vital in piecing together what we knew about outer space. The most well-known, by far, was Allan Hills 84001, the now infamous Martian meteorite that had electrified the science community six years earlier and solidified my desire to unravel the mystery of the origin of life. After the fanfare surrounding that discovery, skeptics began insisting that the sample had been contaminated with many of the purported biosignatures *after* its landing in Antarctica. It was hard to imagine this happening in such a desolate place, but one day, Danny and I came across a meteorite sitting in liquid water below the ice. The persistent sunlight had caused the dark rock to reradiate heat, creating a mini greenhouse effect and melting the ice around it.

"Dude," Danny exclaimed, "it's just sitting in water!" He grew agitated, trying to convince me that terrestrial amino acids could be contaminating the meteorite as we watched, sullying the sample, and rendering future lab findings unreliable. He had been part of the team that studied the amino acids in ALH84001, hoping to find signs of alien organisms. It was a maddening concept—that even in the most pristine environment on Earth, the information carried within could be compromised. It was like finding buried treasure, only to realize that it was all counterfeit. My confidence that we could really unravel the origin of life with these specimens was understandably shaken.

Nevertheless, the samples we retrieved were the most unspoiled on Earth and the work some of the most meaningful I had ever done. Even when the weather turned brutal and it was difficult to handle the plastic bags, the little tags, and the snowmobiles, I yearned to stay in the field. On the worst days, Jamie stepped in and announced that we were heading back to camp early. I was part of the vocal minority in our group that protested, thinking about all the material yet to be found. But when he patiently explained that even the tiniest mishaps could put the expedition in jeopardy, I conceded, following his lead back to camp.

"Tent days" are a part of the Antarctic experience. Katabatic winds are pressure-derived gales produced when cold air sinks over the Antarctic continent and races over the ice toward the ocean. There is nothing to stop the airstream, and the mountains only funnel it. Some mornings, we poked out from our tents to be greeted with windchills below −70°F. On these days, we stayed inside, except to make a quick dash to relieve ourselves when necessary.

I was an avid backpacker, and the tent I shared with Danny quickly began to feel like home. Compared to the lightweight gear I'd used in the desert, the enormous Scott tent was a palace, even with two of us in residence. On the other hand, most palaces have indoor plumbing, a luxury we lacked. For six weeks, I peed into nearby crevasses, thinking about the implications of my DNA locked up in Antarctic ice for millions of years to come. The eight of us also shared what we lovingly called the "poop tent," a petite orange pop-up tent with a bucket, toilet paper, and some hand sanitizer. All solid human waste must be flown back to McMurdo for disposal, and my least favorite task was changing the bucket when it was full.

As far as personal hygiene goes, we bathed with baby wipes. This kept us clean but not sweet smelling, and after a while, we hardly even noticed each other's funk. We worked for our water out there, melting chipped ice on our gas stoves. We depended on these stoves not only for cooking but also for heat. Refueling the stoves—which we had to do often—was a challenge. They were ancient and commonly burst into flame right in the middle of our tent. The first few times this happened, it was frightening. By the end of the campaign, we routinely opened the tube that led out of the tent and calmy chucked the flaming stove out onto the ice to cool off.

The weeks wore on, and the separation from Kate started to weigh me down. We chatted every three days by satellite phone for a brief five minutes. The hardest days were the holidays. For Christmas, she was with her family, surrounded by warmth, good food, and cheer. The dichotomy with our holiday breakfast burritos and coffee was striking. Even though she was supportive, I could hear the worry in her voice. I tried to reassure her, but in truth, the mental stress of the isolation was obvious, and she could sense it. I took strength from her concern and redoubled my commitment to the expedition, knowing that she would be waiting for me in New Zealand in a few short weeks.

In addition to the systematic searches on the blue ice, we also hunted on foot in the moraines. For this I had brought my metal detector, the one I used to scour the Arizona desert. At first, everyone was skeptical about how well it would work. But when I found a meteorite buried beneath snow and rock, they were convinced. We held "human versus machine" contests, where I scanned with the metal detector while everyone else searched by eye. The machine almost always won. I tried not to gloat, but the immense hauls really lifted my spirit.

We hit a concentration of ordinary chondrites on one day, and I found twenty in four hours. On other occasions, I used the metal detector all day only to find one or two specimens. Some areas were loaded with so many different types of "hot rocks," terrestrial samples that set off the detector, that it became impossible to find meteorites among them. There was also ground that contained plenty of meteorites but did not produce a response on the detector for one reason or another. These rocks were found by painstakingly scouring every square inch—often on hands and knees.

Dante in Antarctica in 2002. Credit: Dante Lauretta personal photo.

Even when not searching, I found myself drawn to the moraines. Something about them reminded me of the deserts of my youth. The piercing loneliness, the jagged landscape, the tiny miracles underfoot. And yet, my thirty-two years paled in comparison to the immense age of the Earth, which was put into perspective when I found chunks of petrified wood dotting the rocky ground, ancient relics that reveal Antarctica's past as a once-lush, forested land. At the bottom of the world, I felt the twin pull of fortune and irony, how lucky we were to be alive and conscious in this universe, and how quickly our lives pass compared to the eons of history recorded in the rocks in front of me.

CARBON APART

THE WANDERING CARBON TWIN DRIFTED through the frigid expanse of the solar system, its journey marked by eons of solitude and emptiness. Something embedded in its parent asteroid was resisting the deep cold of space—an energy that emanated from within.

The twin discovered the source of the heat: radioactivity. The same stellar explosion that had flung the two carbon atoms into space had produced radioactive elements, and they were leaking gamma rays. Over millions of years, the heat inside the asteroid had grown, and it had transformed from a barren, lifeless rock into a place of pulsing energy and boiling fluid.

Ice melted and migrated through the asteroid, like an

extraterrestrial hot spring, and as the fluid liquified minerals along its path, the tar holding the carbon dissolved. Three oxygen atoms enclosed it, locking the wandering twin in the center of a carbonate compound.

The carbonate molecule was enraptured by the liquid, eagerly gaining electrons to go with the flow. As it drifted past the primordial rock, it beckoned to other elements, "Join us in solution!" And calcium atoms answered the call, leaving their host rocks and becoming free-floating ions whenever the carbonated liquid came calling. Together, they formed a grand solution that penetrated every crack and crevice in the asteroid.

But like all things in the universe, the radioactivity was limited. After five million years, the asteroidal hot tub party was over, and the wandering carbon atom knew it was time to find a solid place to call home.

It encountered a calcium ion and bonded immediately. As the water cooled, they dropped out of the solution and formed a bright-white salty mineral—calcite veins now cemented the asteroid boulder together.

As the last of the energy ebbed, the wandering carbon atom reflected on its twin. Had it also found a good home? It called out across the solar system, wondering if it would get a response. Nothing but silence answered back, leaving the twin to continue its journey through the vast expanse of space, alone and adrift once more.

CHAPTER 4

DOWN AND IN

ON A CLEAR FEBRUARY MORNING in 2004, almost one year to the day since I had returned from Antarctica, I stared out my office window from the fourth floor of the Gerard P. Kuiper Building, admiring the Santa Catalina Mountains painted orange with sunlight. It was one of those moments when I felt like pinching myself, knowing that just twelve short years ago, I had opened the school newspaper and learned this field even existed. Now, I was a real professor (assistant professor, anyway) of planetary sciences and cosmochemistry in the Lunar and Planetary Laboratory at the University of Arizona. I was a card-carrying member of the Loony Lab and proud of it.

The place practically buzzed with scientific influence, past and present. Kuiper's legacy lived on in every mention of our building's name. Photos from the Ranger missions—which Kuiper rescued after six back-to-back failures to obtain the first close-up views of the Moon—adorned the hallways.

More recently, an LPL team had built the cameras for Mars Pathfinder, a joyful reminder of a cookout in St. Louis seven years earlier. On that Fourth of July, the day after I defended my PhD thesis, a group of us had gathered for brats and beer to watch CNN's coverage of the rover touchdown in the Ares Vallis. We erupted in applause as the first images flickered across the screen, revealing round pebbles and cobbles at the landing site, looking for all the world like tumbled rocks in a dry Arizona riverbed.

Now, teams in my home department were leading investigations using instruments on Cassini—a spacecraft currently en route to orbital insertion around Saturn and its companion lander, Huygens—destined for Titan, the planet's largest moon. Another team was busy building the High Resolution Imaging Science Experiment, a spy camera that would fly on board the Mars Reconnaissance Orbiter. It seemed I couldn't turn a corner in the Kuiper building without overhearing a spacecraft team agonizing over every minute detail of their mission.

Most of these conversations centered around one target—Mars. In 2003, LPL had won a bid to lead the Phoenix Mars Lander, to look for water on the red planet. As significant as this announcement was to the UA community, and Mars scientists in general, it also signaled a big shift in the organization and funding of space exploration.

Phoenix represented NASA's new approach to mission teams. Unlike the massive flagship missions such as Cassini, which were centralized under NASA management and cost billions of dollars, these missions would have smaller budgets and involve collaboration between universities, NASA centers, and the aerospace industry. The Phoenix science leadership and lander operations were LPL's responsibility, while Lockheed Martin would build and test the spacecraft. Meanwhile, the Jet Propulsion Laboratory, a NASA facility in California, would manage the project. The Canadian Space Agency had been brought on board to provide a weather station, including an innovative laser-based atmospheric sensor. The coolest part of the whole concept was that once JPL placed the lander on the surface, they tossed the keys over to Peter Smith, the principal investigator in charge of the entire mission. The day-to-day operations of the lander would be under the control of LPL scientists.

The ominous ring of my office phone snatched me from my reverie. It was my boss, Mike Drake. As the "kid" of the department, my first thought was that I'd done something wrong. I had made some small missteps early on, trivial things like purchasing equipment without approval or using the wrong copier code. I was sure that this was about another such slipup and prepared my apology. Instead, as Mike's distinctive British accent emerged from the earpiece, I could tell he was excited.

"Dante!" he began, clearly glad to have found me in my office. "I've got Lockheed Martin here. They want to partner with us on a new mission and I want to bring you in. Can you meet us at the Arizona Inn after work for drinks?"

I told Mike of course, pretending like this surreal combination of words was totally normal.

Hours later, I walked into the courtyard of the Arizona Inn, the historic boutique hotel northeast of campus, inhaling the scent of its lush gardens. As I stepped inside the patio doors of the Audubon Bar, I located Mike's bald head and big glasses and took a seat in the empty chair beside him.

"This is Steve Price," Mike said, waving his hand in the direction of our companion, "director of business development at Lockheed Martin."

I shook Steve's hand, and we placed our orders. Before our pleasantries were out of the way, the server had returned with a glass of scotch for Steve, another for me, and a bottle of Kendall-Jackson chardonnay for Mike.

Steve got right to the point.

"As you know, we have a steady line of business building interplanetary probes for NASA. And we are thrilled with the recent win from the Phoenix Mars Lander proposal." Steve moved his glass out of the way for this next part, the big reveal: "Now we have set our sights on an asteroid sample-return mission. We think it is a perfect fit for NASA's Discovery Program, which funds small missions of planetary exploration. We want Mike to be the PI and for LPL to take the lead on the science."

"That's why I called you," Mike chimed in. "This project would be a natural extension of your astrobiology research."

Shortly after arriving at LPL in 2001, I focused on securing research grants from NASA. Their astrobiology program welcomed me into the fold with a respectable three-year grant for work on phosphorus.

It was my sulfide-formation work back in St. Louis that had led me to this chemical element. As that research progressed, I had increased the complexity of the metal being corroded by the sulfide vapors. I transitioned from pure iron to iron-nickel alloys, then to natural samples cut from iron meteorites. In this last set of experiments, phosphorus behaved very differently from all the other elements. Instead of bonding with sulfur and forming a sulfide mineral, phosphorus stubbornly resisted corrosion. Surprisingly, it piled up at the metal-sulfide interface, forming a thick layer of iron-phosphide minerals. I didn't know then, but this reaction was my entrée into the astrobiology research that would guide the rest of my career.

Phosphorus is central to life. It creates the backbone of our DNA and RNA, forming the supporting rails that connect the genetic bases into long chains. Phosphorus is also integral in the generation of ATP, the molecular fuel that powers growth and movement for every living thing on the planet. Ultimately, it is part of our very architecture. It is in the phospholipids that make up our cell membranes and the minerals that build our bones and teeth.

In terms of mass, phosphorus is the fifth most important biologic element, after hydrogen, carbon, oxygen, and nitrogen. But where terrestrial life got its phosphorus is a mystery. It is much rarer in nature than the other four elements. There is approximately one phosphorus atom for every 2.8 million hydrogen atoms in the cosmos, every 49 million hydrogen atoms in the oceans, and every 203 hydrogen atoms in bacteria.

Working with my graduate student Matt Pasek, we began with the premise that because phosphorus is much rarer in the environment than in biology, understanding its behavior on the early Earth would give us clues to life's origin. Most research up to that point had focused on the common terrestrial form of the element, a phosphate mineral called apatite. The *-phate* part of the name indicates that the phosphorus is highly oxidized, which renders it chemically inert. When mixed with water, apatite releases only very small amounts of phosphorus. Other scientists had tried heating apatite to high temperatures, combining it with various strange, super-energetic compounds, even experimenting with compounds unknown in nature. But this research had failed to explain where life's phosphorus came from.

Meteorites have several different minerals that contain this element, including many that do not occur naturally at the surface of the Earth. So, Matt and I set out to investigate every one of them. After countless experiments, we discovered that the most important one is a metallic phase, known as schreibersite. Unlike the apatite, it reacted like gangbusters in our experiments. Schreibersite is a phosphide compound that is extremely rare on Earth. The *-phide* suffix indicates that it is chemically reduced, suggesting that it could undergo a vigorous reaction with oxygen. It is ubiquitous in meteorites, especially irons, which are often peppered with schreibersite grains or slivered with pinkish-colored schreibersite veins.

Once we zeroed in on this phase, Matt began running very simple experiments. He mixed schreibersite with room-temperature fresh water. Then he analyzed the liquid mixture using an analytical technique called Nuclear Magnetic Resonance spectroscopy. As soon as the first successful experiment was complete, Matt burst into my office.

"Take a look at all these phosphorus compounds," he said, pointing to a graph on his computer screen. "I can't believe how much chemistry is going on in this simple reaction."

I scanned the list of detected chemicals and zeroed in on one of them immediately. "We made pyrophosphate?" I prompted, pointing my finger at the screen.

"Looks like it," Matt replied.

This was an important discovery. Pyrophosphate is one of the most biochemically useful forms of phosphate. Previous experiments had formed this compound, but at high temperatures or under other extreme conditions, not by simply dissolving a mineral in water.

This could really advance the exogenous delivery hypothesis, I thought, contemplating the idea that meteorites may have delivered the building blocks of life to Earth.

Traditionally, this theory had referred to carbonaceous chondrites delivering water and organic molecules to the early Earth. Iron meteorites are completely unrelated to these rocks. They are fragments of ancient metallic cores, like the giant ball of metal at the center of Earth. Our discovery added a whole new class of material to the exogenous-delivery inventory.

I wasn't the only one that realized the significance of this discovery. A couple months later, an editor at *Discover* magazine called to let me know that our phosphorus work was one of their top one hundred science discoveries of the year. With this recognition, I was convinced that asteroids were the key to understanding the origin of life on Earth and throughout the solar system.

Fingering my glass of scotch as Mike and Steve chattered about the possibilities of an asteroid-sample mission, I thought about my bright young graduate student, the labs I was overseeing, and the exciting work happening in them. I considered the money I had just received from NASA to build a brand-new mass spectroscopy lab. What Mike and Steve were offering was beyond exciting, but it would mean giving up a lot.

I was also concerned about tenure. Most assistant professors are laser-focused on their tenure review, and I was no exception. Mine would take place in two years. Planetary missions don't just happen, and projects like the one we were dreaming up could take decades. They involve exhaustive planning and years-long funding competitions — the vast majority of which end in rejection. A failed proposal was hardly the resume-builder I needed to impress the promotion and tenure committees. Even if we did win, designing, building, testing, and launching the spacecraft would take more than five years alone. It is both a technical and political gauntlet, with the constant risk of Congressional disfavor ending in cancellation, leaving behind nothing but PowerPoint presentations and broken dreams.

Even the missions that made it to the launchpad had a spotty record of success. For all the space missions that had succeeded, there were, it seemed, three others that had failed. After the loss of Mars Observer, NASA began sending smaller, less expensive vehicles to Mars, but results were mixed. In 1999, I had sat in a conference room at ASU with the science team of Mars Polar Lander, once again a horrified onlooker as the spacecraft signal went dead. The descent rockets cut off too early, spoofed by the spacecraft's own heat shield, which it mistook for the Martian surface.

These spacecraft failures had left a mark on my psyche. Not least of which was the vision of Laura sobbing in the hallway at the loss of Mars Observer. In 2001, when the Odyssey spacecraft was hurtling toward Mars, I refused to go to the orbit insertion party, even though my friend and LPL professor Bill Boynton had a gamma-ray spectrometer on board. *Maybe I was the one jinxing all these Mars missions*, I ruminated. Sure enough, without my eyes on it, Mars Odyssey managed to lock into orbit, and soon Bill had hired Kate as a spacecraft engineer and data archivist on the mission, her first foray into the field where she would make her professional home.

In the pros column, NASA was clearly interested in sample return. More than thirty years after the last Apollo spacecraft had brought a cache of moon rocks back to Earth, these missions were making a comeback.

The first interplanetary delivery was underway. Genesis, an uncrewed spacecraft that launched in 2001, was spending two years collecting solar wind samples using four circular metal trays exposed to the elements.

The Stardust comet sample-return mission, which launched in early 1999, had flown by comet Wild 2 just one month prior, where it gathered samples from its coma, the comet's gassy envelope composed of ice and dust. To pull this off, the spacecraft was equipped with a tennis racket–shaped instrument that contained a very special substance called aerogel. A silicone-based solid, aerogel was mostly made of air and provided a soft, spongy landing pad for space dust traveling at tremendous speeds. The real trick of aerogel was allowing those samples to come to a stop while minimally altering them physically or chemically. I was busy planning for the analysis of those samples. Their return was my primary justification to NASA for building the mass spectroscopy lab.

The main reason to say yes to this mission was, quite simply, my science demanded it. In my quest to understand the origin of life, I had run up against the limitations of meteorite research. Recalling Danny's agitation over the water-soaked samples in Antarctica, it was increasingly clear that terrestrial contamination literally eats away at the most important prebiotic compounds in the carbonaceous chondrites. Meteorites are altered by the ejection from their parent asteroid and by atmospheric entry. In addition, these materials are very quickly contaminated, colonized, and consumed by terrestrial microbes.

All in all, it was relatively simple to map the pros and cons of saying yes to Mike's offer. There were all the things I'd be giving up, the classes I wouldn't teach, the labs I wouldn't oversee, the long evenings I wouldn't spend with my wife and our future children.

And, of course, I didn't need to convince anyone of the rewards. In the unlikely event the spacecraft did fly, the attention, the accolades, and the sense of accomplishment would be unparalleled. I felt like I was in the final round of a game show, with so much to lose and one irresistible grand prize still behind the curtain.

But the truth is, while I agonized over this decision, there was one variable that did not quite compute, one motivation I couldn't put into words. It was a feeling, one that I had experienced on so many forays into the desert during my childhood. The curiosity that sent me digging through tailings piles in the desert. The compulsion I felt as I filled out my Space Grant application. The impulse that accompanied me to the frozen ends of the Earth. The inner, unknowable voice that took over when I negotiated with Mike for this job. Now my gut was telling me that if I didn't say yes, my curiosity would never be quenched; my questions would never be answered. Where did we come from? Are we alone in the universe?

And so, over shrimp cocktails and drinks, the three of us sketched out a mission on the embossed surface of a cocktail napkin. Lockheed had a concept, Mike had the management credentials, and I knew the science. We didn't have a specific asteroid in mind or definitive scientific objectives, only the guarantee that a spacecraft could be built, and that Mike and I could eventually get our hands on the stuff of astrobiologists' dreams—pristine pieces of a rare, dark, carbon-rich asteroid. The idea that I could pick one of them and have a sample delivered back to Earth seemed like magic, like we were wizards summoning stones from outer space into our laboratories.

"OK," I said tentatively. "I'm in, but I have a couple of requests. First, I will need some funding to cover my summer salary."

"No problem," Steve replied.

"Second," I continued, looking at Mike more as my partner than my boss, "I would like assurance that this is not going to have a negative impact on my tenure review."

Mike nodded in understanding. "I can't make any guarantees, but your current record is strong enough that I can't foresee any problems. I will make sure to provide my full support."

With that assurance, we agreed on our respective roles and responsibilities. Mike would manage the team, handle NASA, and promote the mission to the scientific community—the "up and out." On the other hand, I would concentrate on the scientific aspects, lead the asteroid exploration, and outline the sample analysis program—the "down and in."

This plan, I figured, would give me the time to hone my tenure dossier and learn everything I could from Mike—the administration, budgeting, and politics—of a high-profile endeavor like this one. With that, we were off.

It was late afternoon when I got home from the Arizona Inn, buzzed less on scotch than pure adrenaline. Kate wasn't home yet, so I sat down in my favorite recliner with a pad of paper and a pen. If I was going to oversee the science on an asteroid sample-return mission, I didn't want to waste any time getting started. I began with the big picture—the top-level mission objectives. To win, this mission would need to span a range of topics to satisfy the different factions of the asteroid community, which was fractured along distinct scientific fault lines.

I relaxed into the conforming contour of the chair and tried to quiet my background thoughts. My mind drifted out to the solar system. I recalled the look, the feel, the smell of the meteorites that I had studied. I conjured images of those rocks still in space, wandering the solar system, witnesses to the past four and half billion years. I wondered if any of them were seeking a chance to make it to the surface of the Earth, to reach my laboratory, to tell the stories that they had recorded in their bodies. I reflected on my work with Dr. DeVito, how we used the chemical fingerprints to transmit messages across the stars. I remembered the inspirational stories from *Omni* magazine, where futuristic adventurers hopped across the solar system, and the harrowing images of comet Shoemaker–Levy 9 wreaking enormous destruction on Jupiter.

Bringing my consciousness back to the moment, I wrote down four words:

ORIGINS
SPECTROSCOPY
RESOURCES
SECURITY

Boiling the mission down to these four critical scientific concepts would focus the team, providing a guide when making tough decisions about the spacecraft design and operations.

The first one, origins, was a given; it was the knowledge all my scientific work had been driving toward. Getting our hands on pristine, ancient asteroid material would tell us more than ever before about our origins. This objective satisfied the cosmochemists, the lab folks like me that built our careers around exquisite analyses of extraterrestrial samples.

The second, spectroscopy, referred to measuring reflected sunlight and emitted heat rays from asteroids to infer the minerals and chemicals on their surfaces. Carbonaceous asteroids are notoriously dark, making them a real challenge to study with telescopes. Getting to know one such asteroid up close and personal with a spacecraft, followed by return of samples to Earth, would provide us with a kind of Rosetta stone that we could use when studying other asteroids via telescope. With this objective, I aimed to garner support from asteroid astronomers—a dedicated group of scientists who spend countless nights on remote mountaintops collecting photons reflected or emitted from distant rocky bodies across the solar system to infer their surface composition.

When I wrote down "resources," I was really letting my geek flag fly, drawing on my longstanding interests in science fiction. The idea of mining asteroids for water, organics, and precious metals was a central theme of in situ resource utilization. This community was a small but energetic group of scientists and forward-thinking entrepreneurs. By flying out to a near-Earth carbonaceous asteroid, mapping its chemistry and mineralogy, and returning a piece to Earth, our mission team would bring back a "spec" sample, demonstrating proof of concept and providing much-needed inspiration for the nascent asteroid-mining industry. I imagined future asteroid miners poring over our data as they conceived of the next wave of extraterrestrial resource extraction in the 2020s and beyond.

"Security" was a buzzword I knew would sell. In the wake of Shoemaker–Levy 9's collision with Jupiter, protecting Earth from a similar catastrophe remained a popular prospect for both NASA and the Congressional funding it relied on. It was also a big part of LPL, providing a steady stream of funding. In 1998, Steve Larson and two astronomy undergraduate students, Tim Spahr and Carl Hergenrother, founded the Catalina Sky Survey, using telescopes originally built by Kuiper in the 1970s. Now, after numerous upgrades, the Catalina Sky Survey was the most productive discoverer of potentially hazardous asteroids and a central fixture of NASA's planetary defense program.

Tapping my pen against the paper, I studied my words. The first four letters—O S R S—leapt off the page. On my bookshelf was a well-worn book on Egyptian mythology. I had practically memorized those stories as a kid, isolated in the desolate Arizona desert, and the myth of Osiris was among my favorites.

The story of Osiris loosely paralleled the science of our nascent space mission. Osiris, in his mythological saga, spread knowledge of agriculture throughout the Nile Delta, making modern civilization possible and, in a very real sense, bringing life to the ancient world. He was revered as a god associated with water and thus with the crops along the Nile valley. In the same way, carbonaceous asteroids were believed to have brought the original water and prebiotic organic molecules to Earth, leading to life that sprouted from the barren planetary surface.

After his death, the other gods resurrected Osiris as the deity of the underworld. Similarly, asteroid impacts likely resulted in large-scale catastrophe and mass extinctions. However, this destruction led to new opportunities and the subsequent rise of new species. Indeed, we probably owe our own origins to the demise of the dinosaurs because of the massive asteroid impact in the Yucatán sixty-five million years ago. It took a minimal leap of imagination to connect how asteroids are like Osiris: bringers of life, harbingers of death.

Suddenly, the acronym appeared. All I had to do was "buy a couple vowels." I slid two skinny *Is* between the letters, and wrote down:

ORIGINS,
SPECTRAL
INTERPRETATION,
RESOURCE
IDENTIFICATION,
SECURITY

and the OSIRIS mission was born.

PART II

CHAPTER 5

WHATEVER IT TAKES

LOCKHEED MARTIN LOVED THE NAME, and Mike did too. We began making regular excursions to Denver, setting up residence in a small conference room in the Lockheed Martin Space Sciences Building. Working quickly and with a small team of Lockheed engineers, we assembled a proposal for NASA in just under five months. Full of enthusiasm and more than a little bravado, we submitted the proposal in July 2004.

It bombed.

NASA ranks mission proposals into four categories, from the coveted Category 1 to the lowly, thanks-but-no-thanks Category 4. We were the latter.

The science goals, NASA said, were sound. But the engineering, the management, and the cost were not. Most critically, the target asteroid we chose was poorly understood. Our decision was primarily based on orbital constraints, which made it an easy object to reach. However, our main scientific objective was to understand the delivery of carbon to the prebiotic Earth. Given our lack of knowledge about the asteroid's composition, there was no guarantee that it was rich in this element, which was critical to our research. After a series of professional successes, this rejection, while not totally unexpected, knocked some wind out of my sails. Part of me wondered if I should just cut my losses and move on.

Compounding my doubt, that fall, the Lockheed Martin–built sample-return capsule from the Genesis mission crash-landed in the Utah desert at 193 miles per hour. Its parachute failed to deploy, shattering its precision-crafted sample-collection plates into thousands of fragments, now embedded in Utah dirt. This was Lockheed's first sample-return mission, and the failure came after the spacecraft had spent two years in space collecting particles of solar wind. Its design would be the basis for the OSIRIS spacecraft. If NASA had been unimpressed with the engineering before Genesis slammed into the hard desert ground, this latest mishap surely wasn't going to help.

But as I vacillated, Mike was resolute. Genesis's crash landing, he insisted, was good news for us.

"How is that possible?" I asked.

"Sometimes failure is just as important as success," he explained at lunch one day, working through his midday bottle of chardonnay. His British accent could make even the most dubious statement sound reasonable. "Now those engineers can take all those lessons they learned and apply them to our spacecraft."

A few days later, I sat in his office, more than a little impressed, as he made the phone call to Lockheed Martin. Like a spurned patron at the customer service counter, he demanded we get access to the best engineers they had to offer.

"We need their A-team; once we get access to the right people, the engineering will fall into place," he told me with a confident smile. "You just worry about the science."

Up and out, I had to remind myself, was Mike's job. Down and in was mine.

Mike was a gamer, a strategist, absolutely determined to get his way. I had already known this before that fateful afternoon at the Arizona Inn. The man was a legend in our field. Intellectually fearless, Mike had made the centerpiece of his career the giant question of how the solar system formed and evolved, performing daring experiments at high pressures to simulate the core formation of planets. At nearly sixty years old, he had already won almost every award and accolade our field had to offer. The OSIRIS mission was going to be the crowning jewel of his trophy case, the final flourish on an already flawless legacy.

I thought a lot about legacy as we pushed forward with the second proposal. At the forefront of my mind were Gerard Kuiper, Carolyn Shoemaker, Bill Boynton, and Phil Christensen, scientists who kept after their dreams, day after day, even when the community scorned them, the system seemed stacked against them, or their spacecraft exploded right before orbit insertion.

I also contemplated the future. If we were successful in

getting pieces of asteroid back to Earth, those samples would inform countless studies, doctoral theses, whole careers. The vault at ASU had changed the trajectory of my life. I wanted future scholars to have the same opportunity. And when Dani DellaGiustina became my Space Grant mentee, that future suddenly had a name and face.

Right off the bat, I saw a lot of myself in Dani. Coming from a run-down neighborhood in El Paso, Dani was a straight-talking, rock-climbing desert rat like me, one who had helped her single mother raise her troubled younger brother. But she was also a polymath, gifted with a free-wheeling intellect that helped her stand out from all the other—very qualified—applicants. Dani's giant brown eyes were seemingly always wide with wonder. It made me remember my own dizzying days as a Space Grant mentee.

Matched with me to work on a research program in asteroid science, Dani wanted to investigate how asteroids might help humans find safe passage to Mars. By focusing on asteroids that fly past both Earth and Mars in a single orbit, she developed the "Martian Bus Schedule." These asteroids, she posited, might serve as radiation shields for astronauts making the perilous journey to the red planet. The work attracted a lot of attention and award recognition.

It was this kind of out-of-the-box thinking that I wanted on the OSIRIS team. Because no one had ever attempted to land on an asteroid, and since there was no list of best practices we could fall back on, we needed creative, forward-thinking scientists like Dani not only to secure our sample, but also to spearhead future expeditions of deep space exploration. She signed on to lead a student experiment for the OSIRIS mission.

I began to see that our burgeoning mission was about so much more than science. It was about people—Mike's legacy and Dani's future. My job in the present became crystalline: Go find the right asteroid, one with a good chance of being carbon rich. Now it was time to decide which one.

✦ ✦ ✦

By the time I met Carl Hergenrother, he already had an asteroid named after him for his work founding the Catalina Sky Survey. A self-taught astronomer, Carl much preferred spending nights alone in a mountaintop observatory to sitting in a lecture hall. During his years at CSS, he'd discovered three comets and countless asteroids. He was dating one of my graduate students, and we had chatted about his work often. He had also taken my class, Principles of Cosmochemistry, showing a keen interest in the subject. So, I knew he would be the perfect person to consult on which asteroids were worth considering as a mission target.

Carl was already seated at a booth when I arrived, a pint of lager on the coaster in front of him. With a boyish face and long gray hair, Carl stood out from the college kids around us. While they talked about where the party was and who was dating whom, I told Carl, as one does, that I needed an asteroid.

Tucking a strand of silver hair behind his ear, Carl didn't flinch. "OK," he said, waiting for me to give him some more details, something that might help him pluck some possible targets from the 500,000 known asteroids in our solar system.

"Near Earth, obviously. Big enough to have some rego-lith, loose rock, and dust on the surface," I explained, taking a gulp of my beer.

"Makes sense," Carl replied. "We are going to want to find something two hundred meters across or more," he said, "if you want it to have regolith."

"How did you come up with that number?" I asked, always curious to know how my colleagues arrived at such precise values.

"We've been learning about a population of fast rota-tors," he continued. "It turns out that most asteroids smaller than two hundred meters are spinning incredibly fast, some rotating more than once per minute. Any asteroid spinning that fast would have flung all its regolith into space."

I leaned back into the booth. "OK great. One last thing—it has to be carbonaceous."

This statement made Carl's eyes narrow a bit, and I stud-ied his face as he ran through his mental inventory of space rocks. This was a demandingly specific request: an asteroid within easy reach of the Earth that was large enough and spun slowly enough for a spacecraft to snatch a sample from its surface, and that also had to be dark and carbon rich to possess some of those building blocks of life.

"I can think of a few that might fit the bill," Carl said. As we finished our drinks, he promised to send me a list by the following Monday.

On Monday morning, an email was waiting in my inbox. It had been sent at 2:00 a.m. It said, simply:

2001 AE2 (T-type)
1999 JU3 (C-type)

1998 KY26 (C-type)
1989 UQ (B-type)

Four potential asteroid targets, listed by their provisional designations and spectral types. The designation is basically a catalog number that records the date and time of discovery. For example, 1999 JU3 was discovered on May 10, 1999. The spectral classes, which classify asteroids based on the way they reflect sunlight, are where the real scientific value lies. I was immediately skeptical about 2001 AE2. This object was the target of the first OSIRIS proposal, the one that the review board criticized. The T-type asteroids were incredibly rare, and little was known about their composition—there was no solid link between their spectral class and the presence of organic molecules, meaning this object would not work for our mission. Carl quickly followed up, confirming my suspicion. AE2 was really bright, reflecting too much sunlight to suggest a carbon-rich surface.

That left the three truly dark objects: two carbonaceous C-types and one blue B-type. All three had surfaces that were very dark, approximately as dark as charcoal. The B-type asteroid 1989 UQ was especially compelling. This type of object was linked to "active asteroids," bodies in the main asteroid belt that act like comets, appearing to release gas and dust. If these bodies were truly like comets, their outburst indicated icy and organic-rich surfaces, exactly the scientific treasure we were after. I printed out the message and drew a thick circle around 1989 UQ.

My B-type dreams were crushed a few hours later. With the list fresh in his hands, our mission designer emailed me with disappointing results. 1989 UQ was out of range for

OSIRIS. The energy cost would be simply too substantial to overcome. There wasn't a rocket or a spacecraft with enough fuel to get us there.

An hour or so later, Carl followed up with another setback. 1998 KY26 was tiny, no more than 130 feet across. Scratch that one off the list.

That left 1999 JU3. Our mission designer got busy, charting a course through the solar system. One of the big engineering weaknesses in our first proposal was the propulsion system. We wanted to use solar-electric propulsion, a relatively new technology that was running into major problems on NASA's Dawn mission to asteroids Vesta and Ceres, currently in development. Instead, we could get to 1999 JU3 using a bipropellant system, mixing fuel and oxidizer to propel the spacecraft through the solar system. "Biprop" was still complicated, but a step down from solar electric. If nothing else materialized, then JU3 was our target, complex propulsion system and all.

A few years after the Shoemaker–Levy 9 episode, Congress had officially directed NASA to search the solar system and identify the comets and asteroids that might pose a hazard to Earth. Suddenly, a topic dismissed as science fiction had a name pulled straight from the genre. "Spaceguard," a term coined by hallowed science fiction author Arthur C. Clarke, was a global coalition of scientists who—like the Shoemakers—began combing the sky for rocks that could kill us.

One of those teams was the Lincoln Near-Earth Asteroid Research project at MIT. On September 11, 1999, scientists in that lab recorded the first known observations of a dark asteroid they labeled 1999 RQ36. On August 29, 2005, asteroid 1999 RQ36 reappeared in the night sky. It would approach within about thirteen lunar distances on September 20, 2005. Given its close approach, the asteroid astronomy community, including our team, turned their full attention to this rock.

The data were astonishing. Radar observations showed that it was about 1,600 feet in diameter and that it rotated once every 4.3 hours. Spectroscopy combined with brightness and radar data showed that this object was a B-type, one of only a few known among the near-Earth population. Due to its size and the proximity of its orbit to that of Earth, the Minor Planet Center classified 1999 RQ36 as a "Potentially Hazardous Asteroid." As the science community analyzed the radar data and refined the orbit, this object shot to the top of the list for planetary defense.

Furthermore, the data were so good that we could resolve this asteroid's shape with astonishing clarity. As the 3D model rotated on the screen in front of me, I marveled at its nearly spherical figure, stunned that such a small body could have such perfect form. Typically, planetary bodies become spheres when they melt, allowing hydrostatic equilibrium to balance gravity.

Could this thing be acting like a fluid? I wondered to myself.

A clear bulge was visible at the equator and a single boulder appeared in the data, sticking out in the southern hemisphere like a pimple. Otherwise, the surface looked smooth, perfect for sampling.

Almost overnight, 1999 RQ36 went from being a relatively unknown asteroid to one of the most well-characterized in history. The object ticked all our science boxes. It was the longed-for B-type, it was the size and had a rotation state we wanted, with a surface that looked smooth and accessible. The wealth of knowledge from our observing campaign made proximity operations planning much more robust. The final piece of the puzzle was accessibility. Would this jewel remain just out of reach, like its cousin 1989 UQ? I sent the target over to the mission design team for analysis.

They responded immediately. "This target looks good. It is in a very Earth-like orbit, the trajectory looks quite doable. What's even better, we can reach it with a monopropellant propulsion system." Monopropellant is the simplest, most tried-and-true kind of propulsion system out there. If we were looking to tighten up our engineering story, then this was a huge step forward. Just like that, we had our target. My face broke into a grin so wide I could feel the muscles ache.

We found you, I thought to myself as the surety of the decision radiated from my core.

Everything about this asteroid felt right.

In early November 2006, two weeks after Kate and I welcomed our first son, Xander, we got the call that OSIRIS was selected by NASA to compete in a Phase A concept study competition, planetary science's equivalent of the championship game. Instead of their usual two, though, NASA had whittled their applicants down to three. We would be competing with GRAIL, a lunar orbiter, and Vesper, a Venus mission. Each team would receive one year and a million dollars to enhance our mission studies.

Even before the announcement came, I could sense we

were making progress. Choosing the most hazardous aster-oid in the solar system gave us a leg up, and in the interven-ing years between our first and second proposals, we had put together a formidable group of scientists and engineers. Mike, I had learned, was a magnet for top talent. While he liked to call himself a "benevolent dictator," the truth was that Mike was a compelling leader. He was a bulldog, a man who thrived on pushback, whether it was from Lockheed Martin execs or members of his team. He never asked any-one to work harder than he did. Watching Mike was like taking a master class in leadership.

While we wrote and rewrote our proposal, sitting next to each other on airplanes, at bars, and in conference rooms, I also began to see beyond his well-honed British veneer to the man who, for instance, loved careening around the desert in a four-wheeler, or as he called it, his "river Jeep." Or whose first stop in town was always the liquor store, so he could have us over for a social hour (which always lasted much longer than an hour) in his hotel room.

Mike also revealed a softer, more paternal side. When Kate and I bought our first home in Tucson, we threw a housewarming party and invited as many colleagues as we could squeeze into the mid-century ranch. After surveying the place, Mike announced it was perfect, then took Kate aside to assure her the department, the university we had yoked our lives to, would take care of us. We were going to do great things, he promised. When Xander was born, Mike was immediately smitten, marveling as our boy learned to crawl in the corridor of a Maryland hotel.

Yes, we were sailing now. Our team's morale was high. Lockheed had stepped up with their finest engineers, and

NASA's Goddard Space Flight Center had come aboard to provide project management, a category in which we had received low scores the last time. And while the other two mission concepts we were competing against in Phase A were sound, they weren't sexy like a landing on an asteroid and scooping up some space rocks.

So, in December 2007, when our second proposal was rejected, it wasn't just stinging, it was shocking. At the debrief, NASA provided a glowing review of the mission and our team. The only major weakness was the details of our work plan, an easy fix. NASA tried to select us, they explained. They even tried saving money by having us share a rocket with GRAIL, the eventual winner. It didn't help that the Dawn mission, with its problematic propulsion system, had experienced a launch delay that cost NASA millions of dollars, money that might have put us over the top. Despite their best efforts, ultimately, OSIRIS was just too expensive, NASA said.

A couple years before our second rejection, NASA had launched the New Frontiers program, an initiative that funded more expensive, more ambitious missions with a high science return. Named after President John F. Kennedy's famous speech in which he argued that America still had "uncharted areas of science and space," to explore, New Frontiers missions came in at double the budget of a Discovery-class mission. NASA had already green-lit New Horizons, a mission to Pluto, and Juno, the first solar-powered spacecraft to explore an outer planet. In 2008, the body in charge of setting the priorities for New Frontiers missions, the National Research Council, announced it would be updating those priorities, possibly opening up the program to new kinds of missions. Mike and I knew that if we could convince the NRC that a carbonaceous asteroid sample-return mission should be among those priorities, our budget problems would evaporate, and OSIRIS would fly.

Once again, I watched Mike get to work, broadcasting the results of our efforts to the scientific community. Working with key team members, Mike secured an appointment to brief the NRC about our mission. Our science was peer-reviewed and judged to be of the highest value to the agency. Our engineering was strong, simple, and capable of pulling off this mission. Our challenge was the budget, the exact resource that New Frontiers unlocked. If any mission deserved to be in this category, it was OSIRIS.

When the NRC published their revised recommendations in 2008, asteroid sample return was on the list. Not only did our budget double, but NASA would also now provide a more capable rocket. The extra dollars and souped-up rocket would allow us to launch something more than double the mass of our current concept.

This sudden influx of resources had my head swimming. In an instant, all the challenges we agonized over were gone. Like a kid writing out his Christmas list, I added more science instruments, more engines, and abundant fuel with which to maneuver around the asteroid. I could practically feel the wind at our back. On the other hand, there were two other very exciting sample-return missions in contention, including one to a comet and another to the lunar farside — stiff competition.

Mike and Dante circa 2007. Credit: University of Arizona Special Collections.

Our first two rejections had taught us a lot (Mike was right about failure), and our Phase A invitation signaled that NASA was into our idea. As far as I was concerned, we were heading into the New Frontiers competition with the lead, but it wouldn't cut it to just coast. We needed to make the mission faster, stronger, better. When the team gathered in Tucson for a kickoff meeting before our third proposal, I took the lectern and said as much. "We are going to win New Frontiers 3," I declared, with the bald confidence of a presidential candidate, "and we are going to bring a piece of asteroid 1999 RQ36 back to Earth."

Faster, stronger, better, it turns out, was a lot easier with the extra budget we would have under the New Frontiers funding line. We added new team members, the Justice League of planetary science, as I began to think of them. We brought in Peter Smith, leader of the Phoenix Mars Lander, Phil Christensen to build one of his trusty thermal emission spectrometers, Carl Hergenrother to head the asteroid astronomy team, and my old Antarctic tentmate Danny Glavin to lead the organics analysis. Bill Boynton joined up, applying his knowledge and battle scars to help us avoid the pitfalls of the past. Mike Nolan, who led the radar observations of 1999 RQ36 at Arecibo Observatory, also came on board.

Each of these experts recommended more folks, and the team grew exponentially to include members with experience on previous missions to Mars, Mercury, Venus, Jupiter, Saturn, and Pluto. World experts from NASA, the Smithsonian, and leading universities focused on the sample science. Dynamicists who study the wandering of asteroids through the solar system started plotting the security investigation. Our online and in-person meetings were a smorgasbord of colorful personalities, some eccentric, all brilliant. It was a significant departure from the first proposal, when our mission staff was a mere handful.

The anticipated science payload was now awesome. Our last mission had been bare-bones, flying only a few cameras to characterize the asteroid, grab the sample, and get home as quickly as possible. Now we were a true asteroid remote-sensing mission. We added three science cameras: the Poly-Cam to find 1999 RQ36 from millions of miles away and perform microscopic imaging of the surface, the MapCam to survey the asteroid in full color, and the SamCam to witness the contact with the asteroid's surface. We bolstered our spectral interpretation science with three new instruments: a visible and infrared spectrometer, which would measure the chemistry and reflectance of the asteroid, a thermal emission spectrometer to sense the heat emanating from the surface, and a laser altimeter from Canada—based on the successful instrument that flew on Phoenix—to rapidly create three-dimensional maps. We also sponsored a student project, the REgolith X-ray Imaging Spectrometer (REXIS), to train the next generation of mission leaders.

The last piece of the puzzle was the mission name. OSIRIS had great brand recognition at NASA Headquarters. However, OSIRIS was a Discovery-class mission. We needed a name that acknowledged our heritage but indicated we were, in fact, bigger, better, and stronger than before. Out of nowhere, a team member suggested OSIRIS Rex, followed by chuckles from around the room. But into the evening, the name continued to resonate in my head; it sounded like a New Frontiers mission. The name's ability to conjure images of a formidable dinosaur only added to its allure; everyone associates these giant reptiles with a devastating asteroid strike. Like my first flash of inspiration, this one came suddenly: a backronym, using two words to make the acronym work: *REGOLITH EXPLORER*. After all, it was 1999 RQ36's regolith that we were after, and like explorers of old, we were in search of treasure and adventure. We were now OSIRIS-REx.

Of course, with more money, scientists, and instruments came the problem of managing all of it. Mike and I both felt like we needed a dedicated team member to wrangle personalities and dollar amounts and keep us on track. Luckily, we only needed to look down the hall to find her.

Heather Enos had started her tenure at UA in 1991 as an accountant in the financial aid office. She quickly rose through the ranks and then into a completely different league of ranks, when Eugene Levy (the man who had dashed my SETI dreams) asked her to help the university manage NASA contracts. She accepted that offer and went on to help Bill Boynton rebuild his lost gamma-ray spectrometer for the Mars Odyssey mission, overseeing the budgets and schedules of engineers, scientists, and scholars. Small in stature, Heather dominated the room. Her knowledge of spaceflight hardware was encyclopedic, and any deficiency was made up for by her painstaking note-taking. Her mission notebooks were so thorough, they are now archived in the UA special collections department.

Heather would go on to be a part of the Phoenix Mars mission, which landed on the Martian surface in 2008. The lander's assignment was to explore the red planet for several months. It had been examining patches of underground water ice that had been detected by Odyssey, hoping to find evidence of the planet's ancient past and its potential to support life.

As the lander dug deeper, it identified something truly astonishing: calcium carbonate minerals, which suggested the occasional presence of thawed water. This was a groundbreaking discovery, as it suggested that Mars might have had a much more hospitable climate in the past than previously thought.

But that wasn't all. The lander also found soil chemistry with significant implications for biology. The data showed that there were organic compounds present, key ingredients for the formation of life as we know it.

However, the mission's biggest surprise came when the lander discovered perchlorate, a chemical on Earth that is food for some microbes and potentially toxic for others. This discovery changed the way we thought about Mars. As researchers pursued the implications of this finding, they discovered that chloromethane and dichloromethane were produced when a little perchlorate was added to desert soil from Chile, the best analog for Mars soil on Earth. These compounds were the very same ones seen in the Viking tests three decades earlier, the ones that had been ruled out as terrestrial contamination. Phoenix's result suggested that these early Mars probes may have found evidence for life after all.

As the Martian sun set on Phoenix, the solar-powered spacecraft shut down, its mission a resounding success. Heather came aboard OSIRIS-REx, bringing her depth of knowledge and experience. With the team in place, we got busy working on yet another proposal to fly an asteroid sample-return mission, the third in four years.

It was 2009, at 6:00 a.m. or so, on one of those languid days between Christmas and New Year's, when the phone started ringing. Assuming it was bad news, I felt my heart beating as I brought the receiver to my ear.

"Dante, this is Tom Morgan." Tom was the New Frontiers program scientist at NASA Headquarters. "I'm glad I reached you. Congratulations—OSIRIS-REx has been selected for Phase A!"

My heart leapt in my chest—back to the championship game—this time, hopefully, we would emerge triumphant.

After a moment of euphoria, my chest tightened as a new worry dawned on me. "Tom, why aren't you calling Mike?"

"Actually, I did call Mike. I couldn't get ahold of him."

That's when I realized: It was bad news, after all.

Mike's alcoholism wasn't a secret. I had smelled the liquor on his breath in morning meetings, shrugged when he ordered one bottle to my one glass. Our colleagues had joked that the best time to ask our PI for permission for something was around 1:00 p.m., when that bottle was well down the hatch. Those hotel happy hours were happy enough, as long as you didn't think too hard about what all the alcohol could do to an overworked, sleep-deprived sixty-year-old man. During the third proposal, the yellow eyes appeared, the sallow skin, the swollen belly.

I had a lot of excuses for saying nothing, for doing nothing. Mike was invincible, or so it seemed. Mike knew what he was doing. Mike wouldn't have listened anyway. But they are just that: excuses. And the morning we learned we had yet again made Phase A, my friend and mentor was in a bed at University Medical Center, the effects of his cirrhosis too dire to ignore. His liver had completely shut down, prohibiting any blood flow. Over the next few days, the doctors installed a stent in his liver and started him on regular dialysis. Slowly, color started to come back into Mike's face.

With our fearless leader somewhat functional again, we prepared for the Phase A kickoff meeting, which would be held at NASA Headquarters in DC. When I landed at Reagan National, the sky over the Potomac was black, the lights of the low-slung city twinkling in the bitter cold. I texted Mike and Heather, who were already in town: *Dinner?*

But as soon as I saw Heather sitting alone in the hotel lobby, I knew something was wrong. Her tearstained face just confirmed it. "Mike's not here," she said. "He's in Tucson in the ICU. He's been intubated. They are not sure if he is going to make it through the night."

With no other option, I walked into NASA Headquarters the next day, sat down at a conference table next to an empty chair where Mike should have been, and pretended everything was going according to plan.

The debrief was led by NASA program scientists. Most of the meeting has faded from my memory now, except for their opening spiel: "The OSIRIS-REx proposal is as close to perfect as any spacecraft mission proposal that NASA Headquarters has ever seen," they said. I stole quick glances with the team members in the room that morning, congratulating each of them with my eyes.

The briefing got even better. First, OSIRIS-REx was classified Category 1. Per NASA policy, "Investigations in Category 1 are recommended for acceptance and normally will be displaced only by other Category 1 investigations." Second, the other two missions selected for Phase A, the MoonRise lunar sample return and the SAGE Venus lander, were classified as Category 2 missions. According to NASA, "Category 2 missions are well conceived and scientifically or technically sound investigations which are recommended for acceptance, but lower priority than Category 1."

The bottom line from the meeting: It was ours to lose.

The biggest threat to our chances of winning the mission was the precarious situation of our leadership team. Two days after the debrief, I sent an email to mission leadership, letting them know that Mike was on medical leave and that I would serve as acting PI. As University leadership urged Mike to retire altogether, I plunged into the daily chaos of running a spacecraft mission. My meetings swung from intricate engineering reviews and trajectory design around the asteroid to dealing with irate team members based on the content of NASA's latest press release. My "down-and-in" days seemed like a pleasant dream cast against the frazzled reality of leading this effort by the seat of my pants. Life was nonstop on-the-job training, and I was taking a crash course in project management and systems engineering, with a dash of science on the side.

I was beginning to think my time in the captain's chair was permanent when Mike was scheduled for a liver transplant in May 2010. The next time I saw him, the yellow pallor was gone from his eyes, which looked sharp and focused for the first time in a long time. A month later, Mike was back in the trenches, stepping in as if he had never left.

I urged him to take it easy. He could handle the big-picture management stuff from his office in Arizona, I argued; no need for long flights to DC or trips to Colorado to visit with Lockheed Martin reps. I could handle all of that. But he'd always wave me off. "I feel great," he'd say. "Plenty of energy."

Mike went hard, taking on a travel schedule that would have been brutal for a young man, let alone an old one with a new liver. It just wasn't in Mike's nature to pull back, especially at what was arguably the mission's most important moment. And Mike nailed every meeting, briefing, and site visit on the schedule, speaking passionately about the

Dante and Mike with the final proposal. Credit: Dante Lauretta personal photo.

impact a mission like ours could have on humanity's understanding of the universe. To make it to the final interview with NASA, Mike endured a red-eye flight from Arizona to DC, taking his seat next to me in the same conference room I had so acutely felt his absence in a year and a half ago. Together, we made one last appeal.

On May 25, 2011, I was sitting on Wrightsville Beach in North Carolina with my family, which now included our second son, Griffin, born two and a half years earlier. With Mike relatively healthy at home in Tucson and NASA still weeks away from their decision, I felt, for the first time in a long time, relaxed. Over the last seven years, we had accomplished an incredible feat, conceptualizing, planning, and perfecting a space mission to the most dangerous asteroid in our solar system. Along the way, I had formed some of the most meaningful relationships of my life. Whatever happened now was out of my hands. As I dug my toes into the cool sand, I knew these details would be small comfort if NASA ultimately rejected our proposal, but some comfort still.

Just then, I felt my cell phone buzz against my leg. It was Mike. NASA had made their decision early, he said, and after seven years of effort, *OSIRIS-REx is going to fly.*

CHAPTER 6

THE MYSTERY OF THE DARK ASTEROID

AT THE NEWS FROM MIKE, I felt light-headed. After years of challenges, duress, and pure grit, we had won. The dream had felt so far out of reach for so long, I almost couldn't believe it was true.

Mike passed the phone to Heather. "We did it!" she screamed. "All those long days paid off! Now, we actually have to build this thing."

Her words shot through me like a triggered nerve, and I felt a sudden surge of panic.

"Holy shit!" I burst out. "We have to build an asteroid-exploring robot." Even after seven years of intense effort, flying the actual mission had always felt theoretical, shrouded in uncertainty. Now that the future was firmly in place, the enormity of what I had signed up for sank in.

Another spike of panic shot through me as I thought, *Is Mike up for this?*

What we didn't know then was that sometime on that last red-eye flight to Washington, DC, as the night of the Southwest gave way to morning on the East Coast, Mike's left lung, weakened from pneumonia, had collapsed. As we presented our final pitch to NASA, fluid expelled from that lung had already begun circulating in his chest, creating a moat around Mike's heart, where it would spend the next few months thickening, making it increasingly difficult for him to eat, to move, to breathe. By September, when his doctors discovered what had happened, his pericardium had hardened; the only option was to surgically remove it.

Mike sent a cheery email saying he expected to recover and be back at work in about a month, explaining that during that time, I would once again be PI. Despite his optimistic outlook, a sinking sensation settled in my stomach. Against my better judgment, I researched the survival statistics for the procedure. The likelihood that Mike would make it through the surgery was a mere 25 percent. My eyes blurred as I read the number, emotions swinging from panic, to anger, to dread.

On the day before his surgery, I visited Mike in the hospital. Surrounded by stacks of old issues of *Science*, he greeted me with an upbeat smile, no sign of pain or worry on his face. Instead of sitting in the cold vinyl chair in the corner, I leaned against the wall at the side of the bed. Unable to make the kind of small talk I knew the situation demanded,

I felt desperate to address the obvious: that he might not make it and that, if he didn't, I wasn't sure how I—or the mission—would carry on without him.

As deftly as I could, I conjured the billion-dollar elephant in the room. "We need to talk about what your surgery means to the mission," I began. "I mean, from a spacecraft project perspective, this is a red risk." In NASA lingo, a red risk is the most severe challenge a mission can face, one that requires outside assistance. "Have we considered any risk mitigation strategies?"

Mike just smiled. "Dante," he said gently, "*you* are the risk mitigation."

If I had hoped to conceal my emotion behind shoptalk or scientific jargon, my mentor had called my bluff. My throat tightened and my fingers trembled. The pain of imagining a future without my friend roared through my chest like a terrible thunder.

"That wasn't the deal," I half joked, letting the tears roll down my face. "My focus was *down* and *in*, remember? I'm the scientist, the lab guy, the cosmochemist. I don't know if I can do this."

To the very end, the PI of OSIRIS-REx was steadfast, serene, impossibly confident. "It's your mission," he told me, knowing I would understand all of the love, trust, and pride wrapped up in those three words.

Mike did end up surviving the surgery, but the prognosis was grim. He would live a few months, the doctors said, though he would likely spend them in the hospital. But just weeks later, in the middle of a meeting at the Goddard Space Flight Center in Maryland, my phone rang. It was Mike's wife, Gail.

"If you want to say goodbye, you need to get to the hospital right now," she said.

"I can't," I muttered helplessly. "I'm two thousand miles away."

Mike passed away on September 21, 2011, leaving an unfillable void in the field of planetary science and in my life. The void he left in the mission, however, would have to be filled—no matter how impossible that seemed. With Mike gone, I was now the sole individual that NASA looked to for mission success.

✦ ✦ ✦

Every spacecraft mission has at least one moment of terror. For most missions, this is orbit insertion or landing. For OSIRIS-REx, this moment would be the five seconds when our spacecraft "kisses" the surface of the asteroid and gathers up a sample.

As I grappled with grief and despair following Mike's death, those five seconds began occupying nearly every second of my time.

"It is easy!" we had blustered in the proposal. "Everything moves so slowly! There is *essentially* no risk!" we implied, trying to convince ourselves as much as the review boards. The proposal was all about bravado.

But selling a concept in a proposal document is a far cry from building a real spacecraft. These missions are marathons, and selection is only the first mile. Phase B, as the next epoch of mission formulation is called, came with millions of dollars in funding. Hundreds of individuals came on board, most of them young engineers eager to solve the challenges of this awesome mission. Our focus became the next milestone, the "confirmation review," which would take place in 2013. We had two years to convince the NASA Administrator and his advisors that our design was solid enough to release the hundreds of millions of dollars we would need to build, test, and launch the spacecraft.

With the title of PI officially in my email signature, I had to turn my attention to filling out the mission team, which would eventually grow to include more than five hundred people. The first task was appointing a deputy PI, the position I had once held. Before embarking on my search for a new partner, I considered the kind of person I wanted to work with. My heart still ached over the loss of Mike. I knew I could never replace him in my life, but I wanted someone I could lean on, someone I could trust. Though I am not sure I knew it at the time, I was looking for a new mentor and, hopefully, someone I would call my friend long after the mission was over.

I found the perfect match in Ed Beshore. Ed was a self-proclaimed "junkyard astronomer" who had made his living first as a software engineer and now as a research scientist in LPL. Ten years my senior, he had packed up all of his belongings in the 1970s and moved from the cornfields of Nebraska to the rugged desert of Tucson to pursue his dreams of spaceflight. At the time, Ed was serving as the PI of the Catalina Sky Survey, the world's most productive asteroid survey. A self-made man who pursued astronomy as a passion, he brought the perfect complement of skill sets. Beyond his impressive resume, I was smitten by his wandering spirit, bold ambitions, and cheery personality. We hit it off instantly and the pain of Mike's loss eased a bit when I brought him on board.

One of Ed's first pieces of advice was to find a lead imaging scientist.

"The cameras are our eyes out there," he said. "We need someone to own every bit of data they generate. Imaging in deep space is not like snapping a picture with your iPhone. Space wreaks havoc on the camera's detector chips, the lack of atmosphere means light bounces off multiple surfaces, causing glare and flares, and 1999 RQ36 is the darkest object in the solar system. The right person can fix all of that before we even set eyes on our target."

He knew what he was talking about. He had worked on Pioneer 10, NASA's original mission to the outer planets. Ed had helped reveal humanity's first scenes of Jupiter, a painstaking process that required them to line up the data from the single-pixel camera as it scanned the disk of the giant planet one line at a time, building up a picture over a couple of hours. Recognizing the wisdom in his advice, I went seeking my next Osirian.

The universe obliged when Dani returned to Tucson to help her mother and her younger brother, who was in serious trouble with law. My former NASA Space Grant mentee had been in Alaska, where she had completed a master's degree in computational physics by studying glacier dynamics in Greenland. Inspired by her early work on the OSIRIS incarnation of our mission, Dani was devising a plan to land the first seismometer on Jupiter's icy moon Europa, which holds the promise of a deep subsurface ocean, possibly filled with alien life-forms.

I was thrilled to have her join OSIRIS-REx. She needed

a job, and I needed an imaging scientist. It was a big role for someone in their midtwenties, but I knew a thing or two about assumptions based on relative youth. Many of the older members protested, claiming that she was too inexperienced, but I went ahead and appointed her to the position, confident that this polymath was up to the task. We fell into our old roles easily. Like siblings, we reveled in besting each other scientifically, confident that at the end of the day, we always had each other's backs.

It's safe to say that at the time of selection, 1999 RQ36 was the most scrutinized, most understood asteroid in history. But as design started on the spacecraft that would take us there and back, the mission began to feel like what it was—preparing for an adventure no one had ever been on, to a place that no one had ever really seen. As the engineers' questions poured in, I realized that despite the unprecedented astronomical dataset, the depth of our ignorance about the asteroid's environment was profound. Fear of the unknown could quickly overwhelm the team's spirit. We needed a quick victory, something to buoy us before we faced those looming engineering challenges.

One area of uncertainty could be quickly resolved. Instead of a cold, soulless catalog number, we needed to make our asteroid into a real place. It needed a name. So, with the help of the Planetary Society, we launched a worldwide *Name That Asteroid!* competition in an attempt to get the public involved in the mission and give our space rock a more memorable moniker. The contest was open to children of school age from all over the world.

More than eight thousand students submitted entries. We sorted through stacks of Greek gods, Norse dwarfs, and

Hindu warriors. One kid suggested we call it Kal-El: Clark Kent on the outside, Superman on the inside. Another recommended Voldemort. For a while, I lobbied hard for Rama, an homage to legendary sci-fi author Arthur C. Clarke and his inspiring work *Rendezvous with Rama*.

The winner was an eight-year-old boy from North Carolina who said that the sample-collecting arm of our spacecraft, along with its solar panels, reminded him of the neck and wings of Bennu, the Egyptian deity often depicted as a gray heron. Bennu was the Egyptian equivalent of the phoenix, associated with life, death, and rebirth, just like asteroids. Ancient Egyptians often used Bennu as a proxy for Osiris, a living symbol of his divine nature. It was the perfect name. With it, we gained some certainty: We were going to Bennu!

Of course, *what* we would find when we arrived there was still a mystery. As an avid backpacker, I viewed the engineering quandaries in terms of loading up my pack for a new adventure—albeit one to an unknowable, fantastical world. Would I need snow boots or swim trunks? A state-of-the-art GPS unit, or would a simple map suffice? Should I bring more than I think I need, attempt to anticipate every eventuality? Such an approach comes at the cost of a heavy load, one that could ultimately lead to a trip-ending injury. A nimbler approach could be the more prudent path.

Clearly, what we needed was a guidebook. I handed Carl Hergenrother and his astronomy team a big assignment: Synthesize everything we know about Bennu. Since the early days of target selection, Carl and I had worked side by side to refine our knowledge of Bennu. We had also bonded over a mutual love of '70s classic rock bands such as Queen and Led Zeppelin. We were both now fathers of young children and would spend the lunch hour swapping stories about our families' respective Dungeons and Dragons campaigns. In short, I trusted him.

As Carl read through the list of engineering demands, he quickly categorized them into three buckets: *doable, modellable*, and *good luck with that*. In the *doable* category were the asteroid's orbit around the Sun, its size, shape, rotation, and spectroscopic properties. In the *modellable* category were the asteroid's mass, density, thermal attributes, and the presence of any dust or comet-like activity. The *good luck* category included all the critical surface characteristics, like its strength, grain size, and compressibility or "sponginess."

"What do you want us to do about the surface?" he asked delicately. "Some of this stuff is simply impossible to measure with telescopes."

"Itokawa," I replied.

Carl nodded in agreement, understanding the significance of this one word.

Humanity's first attempts to reach asteroids were not unlike the race to the South Pole a century prior. These were expeditions that pushed the boundaries—blew through them, really—of mechanical endurance and human exploration. It was a parallel that only offered so much comfort; after all, a lot of men died lonely, painful deaths on the ice fields of Antarctica.

But like Roald Amundsen, who led the first expedition to reach the bottom of the world, the OSIRIS-REx mission had the advantage of predecessors, all the teams that had embarked on cutting-edge space missions before us. And no mission had more wisdom to share than Hayabusa, the world's first asteroid sample-return mission, and its rendezvous with the asteroid Itokawa.

In the years before selection, we observed with rapt fascination as our colleagues at the Japanese Aerospace Exploration Agency, or JAXA, spent months trying to collect some fragments of Itokawa, a peanut-shaped near-Earth asteroid. But one failure after another—first a software problem, then a busted guidance system, followed by a ruptured fuel tank—hampered their progress. In December 2005, Hayabusa crash-landed on Itokawa and the team lost contact with the spacecraft for more than a month.

Miraculously, JAXA was able to restore communication. They were hopeful that some asteroid particles had made it into the sample canister—the lucky result of tumbling along the surface after the crash. Quickly sealing the return capsule, they began the arduous task of righting the spacecraft and getting it back home.

On June 13, 2010, Hayabusa's return capsule was delivered by parachute to a barren patch of the Australian Outback. A few months later, JAXA confirmed that thousands of tiny asteroid particles—less than a milligram all told—were indeed in the canister. This small sample was massively valuable to planetary science—and to our mission. These heroics would be immortalized in not one, not two, but three feature films in Japan. Hayabusa became the Japanese equivalent of NASA's iconic Apollo 13 mission.

Heather's eyes lit up when she learned of the Hayabusa movies. In addition to being a world-class manager, she was like our mission's mom, infusing our scientific work with humanity and camaraderie. Every Christmas, she led the adoption of a family through the Salvation Army, and our office Christmas parties were gift-wrapping sessions, filling everyone with holiday spirit. Sensing the team needed another morale boost, she proclaimed, "OSIRIS-REx Movie Night!"

On a late summer evening, we all gathered at the recently renamed Michael J. Drake Building (now our mission headquarters in Tucson), northwest of the University of Arizona campus in an industrial part of town.

Heather and I did everything we could to make the space into a movie theater, down to printing authentic-looking admission tickets. When they arrived, the team found the auditorium outfitted with futons and couches, suffused with the smell of popcorn. We installed a hot dog stand, a margarita machine, and a folding table with classic theater candy.

Once everyone was armed with snacks and seated, I lowered the lights and cued up the film. For the next two hours, we sat together in a tense and knowing silence as the Japanese blockbuster *Hayabusa: The Long Voyage Home* showed us one version of the adventure ahead.

CHAPTER 7

THE THREE CHALLENGES

WITH ORDERS IN PLACE, THE team's astronomers—a group of two dozen scientists from around the world—got busy doing their best to quantify the values the engineers needed to design the spacecraft. At first, it was madness. Members were emailing each other directly; crossed wires and conflicting messages inflamed tensions. As my inbox filled up with disgruntled missives from both engineers and astronomers, I tried to find some pattern in the chaos, some kind of legend for the map we were trying to draw. As is my wont, I landed on three words and wrote them down:

Deliverability: *Where can we put the space-craft and how accurately can we put it there?*

Safety: *What do we need to do to keep the spacecraft safe?*
Sampleability: *What will the surface be made of and how can we scoop it up?*

CHALLENGE ONE: DELIVERABILITY

At the next meeting, I put the topic of deliverability at the top of the agenda. This element was the most important factor for the guidance engineers.

"As you know," I began, "getting the spacecraft down to the surface of Bennu is one of our most challenging tasks. Where does the spacecraft need to go on the surface? And, most importantly, what kind of guidance system do we need to get it there?"

"Damn right," grumbled one of the engineers in back, "that's a multimillion-dollar decision right there."

I signaled understanding back to him through a quick glance, then raked my fingers through my hair.

"You heard the man," I said. "A lot is riding on this one. Our primary objective is to collect critical data for sample-site selection at exactly the right lighting conditions. Unfortunately," I continued, "for most of the time that we are at Bennu, OSIRIS-REx will be in our so-called 'Safe-Home orbit' where the lighting conditions suck. This orbit is, by definition, the most stable configuration possible. Since Bennu's gravity is so small, the pressure of sunlight hitting the spacecraft is significant, buffeting it like a ship in a storm. The Safe-Home orbit balances gravity and sunlight by following the 'terminator,' the boundary between day and night. The spacecraft can stay in this orbit almost indefinitely, even if contact with Earth is lost for an extended period.

"The only downside of the Safe-Home orbit," I elaborated, "is that from this perspective we are looking at dawn and dusk, with long shadows and dim illumination. We'll leave this orbit from time to time during our adventure. However, these sorties will be brief excursions, and we will always return home to orbit while we analyze the data on the ground.

"Entering such an orbit requires knowing the asteroid's mass. Up until now, this information would have been impossible to measure using telescopes."

I paused to let the tension thicken before I delivered some good news.

"However, we have invented a way to do just that."

What we had done, specifically, was identify the tiny drift that happens as Bennu absorbs sunlight, holds on to it for a little while, then reemits it as heat during its afternoon. In other words, we had unlocked the mystery of the Yarkovsky effect.

The Yarkovsky effect was named after a nineteenth-century Russian engineer who first posited that the orbit of a small, rocky body would change, ever so slightly over time, thanks to sunlight. The idea became widely accepted, but the value of the Yarkovsky effect remained essentially unmeasurable because it is so infinitesimally small. By using radar tracking, we were able to determine that our asteroid had drifted from its orbit by about one hundred miles between its discovery in 1999 and our latest observations in 2011. This distance may sound sizeable until you remember that space is huge. Bennu was 1.4 million miles from

Earth when it was discovered. Tracking its position to this precision is like measuring the distance between New York City and Los Angeles within two inches. The hundred-mile drift meant that the asteroid was slowing down as it hurtled through the solar system, experiencing deceleration.

Yarkovsky predicted that such change in speed resulted when sunlight heats an asteroid's surface, which later releases that energy back into space as heat. So, we needed to measure the amount of heat radiating from Bennu. For that, we booked time on the Spitzer Space Telescope, a mission in NASA's Great Observatories program. Spitzer could detect infrared radiation, which is how heat is emitted from the asteroid. When the photons leave the surface of Bennu, they act like tiny cannons, each one imparting a recoil force.

At this point, all we had to do was invoke Newton's trusty second law of motion: divide that force by the deceleration of the tiny drift, and we could determine Bennu's mass. It was a historical achievement in planetary astronomy — the first use of the Yarkovsky effect to measure the mass of an asteroid. The result was also a godsend to our mission designers, who busily set out establishing our Safe-Home.

The next thing the engineers wanted to know was how accurately they needed to deliver the spacecraft to the surface. Did we need a pinpoint landing, or was the entire surface available for sampling? "Unfortunately," I admitted, "we don't have an astronomy slam dunk for this one. But all hope is not lost. I think, actually, that there is some reason for optimism. For this challenge, our knowledge of Itokawa is invaluable."

Though different in shape and composition, the two asteroids were similar in size. The views of Itokawa returned by Hayabusa revealed that, even though most of the surface contained abundant boulders, there were fine-grained "seas" made of particles only a few inches in diameter. Plus, the spacecraft had returned specks of dust hundreds of times smaller than what the cameras could detect, proving that sampleable material existed.

"When in doubt, look to Itokawa," I stated. "We are going to use the broadest patch of regolith on Itokawa—about fifty meters across—to define the size of our target zone. Build OSIRIS-REx to sample that."

"Can you give that to me in English?" the guidance engineer asked.

"Really?" I replied. "You're still using imperial units—even after the Mars fiasco?"

I was referring to the Mars Climate Orbiter. A little over a decade ago, I had watched once again as a spacecraft flew directly into the planet's rocky surface instead of entering orbit. Astoundingly, the multimillion-dollar spacecraft was lost on account of an elementary math error. When constructing the navigation system, Lockheed Martin, who provided engineering for the mission, used imperial units of measurement, while the operations team at the Jet Propulsion Lab used metric. No one ever made the conversion.

"I'm afraid so," he replied.

"It's about 160 feet," I told him, likely with a sour look on my face. I really hated working in these units.

"Perfect," he said and smiled. "It should be no problem developing a strategy for guiding the spacecraft down onto such a wide landing strip."

Getting from Safe-Home to the asteroid's surface would require us to fire our rocket engines three different times:

once to leave orbit, known as the "deorbit burn"; once to start the spacecraft's descent toward the surface, known as the "CheckPoint"; and one more time at a "MatchPoint," when the spacecraft would synchronize its motion with the rotation of Bennu, basically making it hover over the sample site, allowing the minuscule gravity to pull it in for a gentle contact.

At the time of sampling, Bennu would be on the other side of the solar system, over two hundred million miles away. My dream of joysticking the spacecraft down to the surface, with a button to fire the lasers, had to be abandoned. It would take eighteen minutes for our radio signal to reach Bennu and another eighteen minutes for us to hear back. OSIRIS-REx would be on its own during sampling.

A few weeks later, the mission designers gathered in the conference room next door to my office to present their proposal.

"Our original idea, the one in our concept study, was to plan each of these three maneuvers in advance, then let the spacecraft fire its engines in sequence." The mission design leader was telling me something I already knew. "Point and shoot," we had called it, telling NASA it would be simple and safe.

"First, the bad news," he said. "As we dug into the concept, we realized that OSIRIS-REx is going to get knocked around by dozens of different tiny forces. We knew about solar radiation pressure, the force imparted when sunlight bounces off the spacecraft, but the truth is there is a lot more to worry about. Since Bennu's gravity is so small, other factors affect the spacecraft's orbit."

He began to rattle them off.

"Bennu is not a spherical body. It has this bulge along its equator, which means the gravity field changes as we orbit.

"Not only that, but its surface reflects sunlight, which hits the spacecraft from below, pushing us away from the asteroid.

"Both Bennu and OSIRIS-REx are hot, relative to space. They both emit heat, imparting force to the spacecraft, which changes our orbit, like a mini Yarkovsky effect from two directions.

"Finally, the nuances of the gravity fields of the Sun and the planets, especially Jupiter and the Earth, play a role in where the spacecraft ends up.

"The good news," he concluded, "is that Pluto isn't a factor."

I stared at him in stunned silence. "What's the bottom line?" I asked hesitantly.

"Point and shoot," he said flatly, "is not going to work. We simply don't know where the spacecraft will be when we leave orbit. We could literally be on the wrong side of the asteroid. We need to add a guidance system to the spacecraft, an autopilot to steer it down to the surface."

The spacecraft needed to get smarter.

From there, OSIRIS-REx would need a way to know where it was after the deorbit burn. Most of the uncertainty on the landing location had to do with the how, when, and where of that maneuver. Given the generous guidance requirement, we would need to program a simple onboard algorithm that would use information from our laser altimeter. This device is basically a laser pointer that measures the time for a single shot to reflect off Bennu and return to the spacecraft. Since we know the speed of light, we can calculate the distance between the spacecraft and the surface.

"Basically," he summarized, "we want to play laser tag with Bennu."

CHALLENGE TWO: SAFETY

Now that we had a way to guide OSIRIS-REx down to the surface, we had to make sure that it would be safe near the asteroid, especially when we sent it in to collect the sample. At the next meeting, we discussed the catalog of questions we had about Bennu's immediate environment. At the top of the list of safety concerns were any particles, either in orbit or as projectiles, that might bonk the spacecraft, damaging it or taking out a critical piece of hardware.

According to Carl, projectiles were certainly possible. He reminded me of Phaethon, a near-Earth asteroid that sheds particles, some of which enter Earth's atmosphere every year, producing the Geminids meteor shower.

Phaethon was a B-type asteroid just like Bennu. The fact that it was an "active asteroid," meaning that it showed comet-like activity, was part of the compelling science. Now that I was looking at it from the engineering perspective, the concept of a rock comet took on a menacing aspect.

"But Bennu's orbit is really different," Carl reassured me. "Phaethon is a sungrazer, meaning it travels deep into the inner solar system during its orbit, getting three times closer to the Sun than Mercury before swinging back out and crossing orbits with Earth."

Carl's confidence, combined with the Spitzer Space Telescope observations, which did not detect any dust around Bennu, allayed some concern about spaceborne debris the spacecraft might have to dodge while at the asteroid.

"Still…" I paused. "The similarities with Phaethon make me nervous. We need to search for satellites and dust plumes on our way in to Bennu. Start developing an observation campaign for the approach phase. At the first sign of any danger, we can apply the brakes and plan a new route in."

After making sure the space was clear for landing, we also needed to identify any tripping hazards on Bennu's surface. With our hard-won measurements of Bennu's mass, shape, and rotation, we put together a map of the terrain that showed all the steep slopes, mountaintops, and valley floors on the asteroid, at the scale of the radar beam, which was about twenty-five feet across, conveniently 15 percent the size of our target zone that the engineers had happily signed up for.

According to our new map, the ideal places to pick up a sample were the valley floors, which, just like on Earth, are where loose sediment ends up. On Bennu, due to the wonky microgravity environment, the longest valley was a belt that spanned the entire equator. Like a Tilt-A-Whirl ride at the carnival, Bennu's rotation almost cancels out gravity in this region, meaning that loose material should be sliding down from the north and south poles and accumulating there. If we were right, there was a mile-long runway circumnavigating the asteroid, waiting for our spacecraft to swoop in and grab a sample.

"The final safety consideration is temperature," I told the engineers. "We have to worry about overheating the spacecraft. Bennu is dark, with a surface blacker than asphalt."

Bennu's day is 4.3 hours long. It experiences a wild two-hundred-degree temperature swing within this short period. Such rapid thermal cycling presents an insurmountable engineering challenge for the spacecraft's most sensitive equipment.

"We can't spend long on the surface. To do so risks overheating followed by freezing the spacecraft. We need to get in, grab the sample, and get out. Borrowing a phrase from aircraft carrier pilots, we are going to perform a 'touch-and-go,' a T-A-G.

"So, we are not just going to play laser tag with Bennu," I told them. "We are going to play laser TAG!"

CHALLENGE THREE: SAMPLEABILITY

"There is one last chapter to write in our guidebook," I began the next week when the engineers gathered once again. "At this point, we have some clarity about where we can 'TAG' Bennu and keep OSIRIS-REx safe in the meantime. Now, we need to figure out what the surface is made of and how we can mobilize some of it into a sample canister in five seconds." In short, we needed to know the size of the rocks that make up Bennu's surface.

"We have a hint in the radar data, which suggest a smooth exterior, with grain sizes substantially smaller than what we see on Itokawa. However, we can't rely on one technique alone. Fortunately, this interpretation is supported by thermal infrared data as well as impressive work from the astronomy team."

Imagine a sandy beach. When you get up in the morning, both the sand and the nearby boulders are nice and cool. As the sun rises, the sand heats up quickly. If you don't bring your flip-flops, you have to make a mad dash to avoid burning your feet. The boulders, on the other hand, heat up slowly. A good strategy to avoid foot burn is head to the boulders and hop along them to safety and, hopefully, a cold beer.

When you go out after sunset, the sand is nice and cool. When you sit on the boulder, however, the heat is still radiating from its day in the sun. The sand heats up and cools off very quickly. It has a low thermal inertia, meaning it does not put up much resistance to a change in temperature. The boulders, on the other hand, heat up and cool off much more slowly. They have a high thermal inertia, resisting changes in temperature.

"The data from the Spitzer Space Telescope gave us some valuable information about Bennu's temperature changes, which are directly related to the size of the grains on its surface," I explained. "It is pretty clear that Bennu has low thermal inertia, meaning that there are most likely fine sand-like particles covering the surface, especially around the equator." At this, I got some nods of agreement and hopefulness; sampling Bennu should be easy!

"To make the case even stronger, we also have thermal data from Itokawa. It has a higher thermal inertia and heats up and cools off more slowly than Bennu. That is exactly what we anticipate based on its boulder-rich surface."

Pausing to see if any of the engineers had questions, I readied my final directive.

"We are going to play laser TAG on Bennu Beach. Pack your bags and get ready for the vacation of your lives."

All of these meetings and data fed into the design of what

was undoubtedly the star of the OSIRIS-REx spacecraft: the Touch-and-Go Sample-Acquisition Mechanism, or TAG-SAM. This proboscis-like instrument would determine how our mission's moment of terror would ultimately play out.

The TAGSAM design had two major components: a robotic arm and a head attached to it. The arm would be responsible for positioning the head for collection, while the head itself would work like a reverse vacuum cleaner. Instead of creating an area of low pressure to suck in air and dirt like a vacuum cleaner does, the head would blow gas onto Bennu's surface to stir up dust particles and collect them. Since Bennu has no atmosphere, we had to bring our own gas to facilitate the sample collection.

Mounted on the forearm would be three bottles of high-pressure nitrogen, a gas that is chemically unreactive, meaning it should not change the sample chemistry in any way. Once we contacted the surface, one of these gas bottles would open up, releasing a jet that creates a high-pressure bubble underneath the head. This bubble would expand upward, pushing rocks and dust like a leaf blower into the head, which is essentially an air filter, much like the ones placed on top of carburetors in classic cars. The arm would then place the head in front of our camera and, if we saw plenty of space goodies, lock the head in a return capsule, like a boot clicking into a ski, and sever the head from the arm. The capsule, with a heat shield that made it look like a mini version of the capsules used to bring astronauts back from space, would then snap shut like a clamshell. Our promise to NASA was that we would deliver roughly two ounces of regolith from the surface of Bennu to the Earth.

Bennu is a relatively small body, but one that still exerts a small pull—its own gravity—on the rocks and dirt it's made of. All our testing was done at the surface of the Earth. Without knowing the exact effect of that gravity, it was difficult to know how the surface of the asteroid would react when TAGSAM made contact—would the different gravity change the amount of sample that ended up in our container? Would larger rocks be sent flying and strike our spacecraft?

In order to figure out the answers to these questions, we needed to simulate the sampling maneuver in an environment as similar as possible to Bennu's. And for that, we booked a week on the Vomit Comet.

Since 1957, NASA's Reduced Gravity Program had been using aircraft to provide astronauts, researchers, and movie directors with the closest thing you can get to a zero-g experience without leaving Earth. For our testing, the effect was created by flying a C-9 airplane in giant parabolas across the sky, reminiscent of cresting the first big hill—and first big drop—on a roller coaster. Near the top of that hill, for about twenty seconds, everything in the airplane becomes weightless. Of course, on the way back up, anything or anyone on the plane must pay back those twenty seconds of weightlessness with a comparable duration spent at two gs, twice the normal pull of gravity at the surface of the Earth, as the plane bottoms out and then begins to ascend toward the next climax. Each flight goes through a series of these "parabolas," every one of them an opportunity to study the effect of microgravity. Unsurprisingly, people get pretty sick aboard these hours-long flights, hence the nausea-related nickname.

In October 2012, I flew to Houston to meet up with colleagues from Lockheed Martin and NASA to test out TAGSAM in microgravity. Bill Boynton accompanied me. Bill was now my mission instrument scientist, a key advisor. His experience base was deep, going all the way back to Mars Observer, that ill-fated orbiter that was lost when I first arrived in St. Louis nineteen years earlier. Since he was always up for an adventure, I knew Bill would enjoy this particular ride.

As we turned into the airfield, a KC-135 aircraft greeted us, appearing frozen at the moment of takeoff. More commonly known as the "Weightless Wonder," this craft flew more than 58,000 parabolas before it was retired in 1995. It was now on permanent display at Ellington Field, a testament to the explorers who had probed microgravity before us.

Bill and I stopped to snap photos of each other in front of the Reduced Gravity Office sign, adorned with the likeness of the Vomit Comet itself hugged by a red parabola, before heading to Hangar 993, where our experiments were waiting for us in giant wooden crates.

We had five flights on the docket, one on the first day and two per day after that. We would perform five tests on each flight, for a total of twenty-five.

We hauled our equipment over to the hangar next door. The Vomit Comet dominated the space, gleaming white with a NASA logo emblazoned on its tail. I stopped to take in the moment, appreciating the rarity of this experience and how fortunate I was to be here for it.

We loaded the five test chambers into the plane. Each one was a square metal box with viewing windows mounted on two sides. A copy of the TAGSAM sample head hung from the top of each chamber, attached to a short piston that would push the sampler into the regolith during each experiment. A tube connected to the side of each chamber and a vent port on the flank of the aircraft would reduce the air pressure in each compartment, so that it matched the pressure outside the plane. It wasn't exactly the vacuum of space, but at 25 percent of the atmosphere at the surface of the planet, it would give us a sense of how TAG-SAM would perform as the gas pressure decreased. Plumbing lines connected bottles of nitrogen with each head,

TAGSAM testing on the Vomit Comet. Credit: NASA.

and valves were automated to open as soon as the collector made contact, directing gas into the regolith then upward into TAGSAM.

After dumping our bucket of gravel into the chambers, I got busy tightening the screws that isolated the experiment from the plane's interior. The last thing we wanted was for gravel to end up free in the plane. If there was one lesson that stuck with me from the safety briefings, it was *don't let anything get loose*. Flying objects have been known to appear mysteriously during the micro-g window, potentially crashing down on someone's head as the plane arced through the bottom side of each parabola.

With the experiments in place, we gathered for a final briefing on logistics, flight conditions, and a run-through of our plans. Our test conditions were demanding, much more so than for most of NASA's passengers. The majority of researchers aboard the Vomit Comet come to study effects at zero-g, ideally with no acceleration at all. However, planes are bumpy beasts, so when pilots hit the zero-g condition, there are small vibrations that cause both positive and negative acceleration. That meant that even in a test when we didn't fire any gas, TAGSAM would come back with a sample inside, thanks to the negative gs that would push material up into the filter.

For our tests to work, we needed the smallest g possible without any negative acceleration. The pilots agreed to target accelerations of 50 milli-g, with vibration adding or subtracting roughly 20 milli-g depending on the atmospheric conditions. The plan was to fly a series of fifteen parabolas per flight, each lasting approximately twenty seconds. The pilot would enter two test parabolas to ensure that our acceleration requirements were met, and then TAGSAM would be fired on the third.

I was secretly elated with this test schedule. Since two out of every three parabolas were test runs, that meant I was free to do whatever I wanted during those periods. I brimmed with excitement at the idea of ten weightless play sessions per flight.

As everyone gave their final go, we headed over to the medical office, where a NASA flight surgeon, who accompanies each flight, supplied us each with a dose of antinausea medicine. "If Dramamine is a water pistol," he said, as he lifted the needle, "this is an assault rifle." I didn't ask any questions, just looked away as he injected the medicine.

When it was time to board, we took our places in the few rows of standard airline passenger seats at the back of the plane for takeoff and landing. As the plane headed out toward the Gulf of Mexico, into the airspace dedicated to test flights, I watched the sky go from gray to white to blue. Once we were airborne, the flight director came over and pulled the window shade down. "You don't want to look out there when we are flying the parabolas. That is a surefire route to puking. And I've seen enough experiments on the behavior of vomit in microgravity," he deadpanned. His experience and good nature beamed from his face as he announced, "Parabola one, coming up!"

I unbuckled my seat belt and headed toward the open section of the plane.

Then I was floating. It felt like cresting the top of a hill on a lonely Arizona highway at too high a speed, when your stomach rises into your chest for that brief instant. Except, the feeling stayed. My stomach kept rising. I found myself in the middle of the airplane, out of reach of any walls, floor, ceiling. A moment of panic shot through me. Foolishly, I started doggy paddling in midair, unable to provide any locomotion to move toward a stable surface. I was trapped.

I caught the comforting blue of the NASA flight suit out of the corner of my eye. Shaking his head, the flight director grabbed hold of me and guided me over to a nearby handle. After twenty seconds, I settled down onto the floor of the plane, gravity slowly turning back on.

"Don't flail around like that," the flight director said over the roar of the engine. "It doesn't help, and quite honestly, you look silly. Find a good launching spot and plan your trajectory. You can fly from the front of the plane to the back in one shot. Make sure that you have a place to land and stay well clear of the experiments. If you want, I can show you how to somersault on the next arc."

"Yes, please," I responded, shamed, but—based on the twinkle in the flight director's eye—I wasn't the first newbie to flounder around in his plane.

Gravity kept growing and I could tell we were heading toward the bottom of the arc. The pressure mounted and I entered a yogic child's pose to ride out the heavy pull. After twenty seconds stuck to the floor, gravity returned to normal.

The pilot's voice came over the intercom. "That was pretty much perfect," he called out, "right down the centerline at 50 milli-gs."

On the next arc, I started to get the hang of it. Watching the flight director, I learned how to somersault. I floated around and snapped a couple of selfies, before handing my camera to Bill for better shots.

Dante flying in microgravity. Credit: Dante Lauretta personal photo.

As we entered parabola three, it was time to watch TAG-SAM in action. We rolled back a protective screen on the regolith bed, intended to keep the rocks in place during the previous two arcs. As gravity receded, I stared at the pile of stones. They didn't move—perfect! The test engineers started the experiment and TAGSAM slowly descended toward the surface. A few flakes flipped up upon contact. They tumbled in slow motion, hovering over the rest of the gravel bed. I marveled at their weightless dance, like a snow globe in outer space.

The bottle fired and I heard the hiss of nitrogen gas expelling from the orifice that ran along the base of

TAGSAM. A tiny dust cloud emanated from the fine mesh that comprised the air filter. The head sank a short distance into the regolith, less than a fraction of an inch. I heard the distinctive rattle of rocks entering the collection chamber. Five seconds later, the gas was gone and TAGSAM rose back up toward the top of the chamber. I examined the surface that had moments ago lain underneath our sample collector. A small dimple was apparent in the gravel, a sure sign that TAGSAM had collected—something.

Twelve parabolas and four tests later, we headed back to the hangar, anxious to see how our instrument had performed. We quickly set about unbolting the test chambers to get at the TAGSAM units within. Lifting each one like a newborn baby, we carried them over to the long tables, lined with plastic to ensure every grain in the filter was collected. I tiptoed behind the test engineer, trying my best not to let my impatience interfere with their delicate work. With the last screw extracted, we held our breath as he lifted the top off the head. There, in full glory, was a ring of black gravel.

I imagined the day, eleven years from now, when I might stare at a similar-looking sample, actual particles of Bennu.

Snapping back to reality, I realized there was one important question I still needed to ask.

"How much did we get?"

As the numbers on the scale settled down, I read the pale blue numbers out loud: twenty ounces. Ten times what we promised NASA.

If our experiments were any indication, TAGSAM would perform flawlessly at Bennu.

INTERLUDE

CARBON ALIVE

AFTER LYING DORMANT FOR FIVE hundred million years, the terrestrial carbon twin awoke. Locked in its own ball of tar, the atom had been content to exist in this eternal state. But now, the environment was changing, and the twin could feel it. Widespread volcanism and frequent asteroid impacts had brought new energy to the planet, tearing apart the tar and releasing the carbon atom.

Suddenly on the move, the carbon atom found more carbon and nitrogen. Together they bonded into a ring, creating a molecule the Earth had never known before—a nucleobase—the first letter of a future genetic alphabet. The joyous ring swirled and spun in the boiling fluids, vibrating with a newfound sense of self. Other rings of carbon formed in the swirling bath, creating rings of sweet sugar.

The nucleobases became enamored with the sugar molecules, and whenever this hydrothermal spring brought them together, they resonated. But one element was missing, and they couldn't overcome the chemical gap that kept them apart.

Then, a massive hunk of metal crashed through the atmosphere, leaving a billowing trail of fiery iron smoke in its wake. It was formed in the core of a protoplanet, where it gathered all the metal together. As the rocks exploded and millions of pieces of metal rained down into the oceans, the missing element emerged—phosphorus.

Overjoyed to be free of its metallic mass, the phosphorus offered a solution. "Join me," it called to the rings. "Together, we can grow into infinity." The rings heeded the call, and phosphorus bonded with oxygen to form phosphate bridges between each sugar, leaving the nucleobases exposed to the outer chemical world, the dawn of the genetic code.

The bridges could link together endlessly, forming chains of nucleic acids that were millions of molecules long. And as they worked together to self-organize, replicate, and evolve, the terrestrial twin came *alive*. It was a moment of pure magic, a convergence of elements and energy that shaped the course of the universe for eons to come.

CHAPTER 8

UP AND OUT—AFTER ALL

ON FEBRUARY 15, 2013, A date that would come to be known as Asteroid Friday, NASA asked me to spend the day speaking with media outlets around the world about Duende, an asteroid then known by its provisional designation as 2012 DA14. That afternoon, at precisely 3:25 p.m. eastern standard time, Duende was going to zoom past the Earth at a distance below weather satellites at 17,450 miles per hour. The event would offer scientists a chance to study a near-Earth object up close—and our team a chance to introduce OSIRIS-REx to the general public.

The day before the asteroid's arrival, I found myself at a conference table at Goddard Space Flight Center with a dozen pairs of nervous eyes trained on me. NASA's media experts seemed to be worried that this young, untested scientist would go on live television and embarrass the agency. With a composed smile, I tried my best to telegraph confidence, but inside, I was surely more nervous than they were. I thought about the kid I had been twenty years ago, sitting on that patio outside the diner, eyes wide at a newspaper ad that said WORK FOR NASA in big block letters. If only he could see me now, preparing to represent those four letters on television.

For the first few hours, we went over the basics of the asteroid's trajectory: It would pass within 20,000 miles of the Earth's surface around midafternoon East Coast time, but only be visible from the Eastern hemisphere, and with the aid of a telescope. As Duende approached and departed our planet, hundreds of observatories around the world would point their lenses at it, generating data that scientists would use to determine the asteroid's shape, composition, and trajectory. If I explained everything right, the event would serve as the perfect media primer for the OSIRIS-REx mission, and I was ready to fill the airwaves with the excitement of the science journey ahead of us.

That's what I reasoned anyway; but as we moved into the cavernous dark of the television studio for on-camera practice, it became clear that I was expected to stick to my talking points with little wiggle room for the kind of compelling commentary I had in mind. As NASA media staff lobbed practice questions my way, I rehearsed my answers, taking care to emphasize the three key messages they had trained me to provide:

1. The asteroid is *not* going to hit the Earth.
2. We are *safe*. The Earth is *safe*. Our astronauts are *safe*. Our satellites are *safe*.
3. I am *excited*. This is a great day for asteroid science.
4. If a reporter asked a follow-up, the bonus message was:

   The prediction of 2012 DA14's flyby is a testament to NASA's *understanding of near-Earth objects* and its ability to foresee asteroid impacts well into the future.

The media trainers were clear: under no circumstances—no matter what question a journalist or broadcaster might ask—was I to deviate from this script. As an example, we reviewed a clip of President Bill Clinton denying his affair with an intern, a masterful performance of evasion. At no point in that interview did he say anything except

Dante on camera during Asteroid Friday. Credit: NASA.

his three key messages, despite an onslaught of aggressive and direct questioning. I had to remain just as steadfast, to avoid creating any kind of public panic about an impending asteroid impact.

Prepared to the point of exhaustion, I went back to the hotel that night and fell into a dreamless sleep. My alarm woke me at 4:00 a.m. and I instinctively reached for my phone to check my email. A message with the subject line "BIG Fall in Russia" was waiting in my inbox from Harold Connolly, a mission scientist and old friend.

Huh, I said to myself. *What a coincidence that a meteorite would fall the same day I'm going on television to talk about asteroids.*

It was surprising, but not statistically improbable. Such falls are observed every few months, some as single stones, others as thousands of fragments strewn along what scientists call a "landing ellipse." As I started getting ready, I thought about the meteorite dealers I knew, all of them clamoring for a plane ticket to Russia at that very moment.

My live shots began at 6:00 a.m., and almost immediately I sensed something was very wrong.

Perched on my swivel chair a few inches from a camera in which I could only see a reflection of my face, the faraway journalist in my earpiece said something about "an asteroid hitting the Earth."

A little confused but undeterred, I rolled out my rehearsed answer about the safety of the Earth, the people, and our satellites, before moving on to how NASA has the ability to foresee asteroid impacts well into the future.

But during the next interview, I heard the same disconcerting phrase: "hit the Earth."

"The asteroid will *not* hit the Earth," I repeated, but the person in my earpiece insisted that it already happened. My heartbeat ratcheted up as I recalled the nervous eyes in yesterday's conference room—was I screwing this up?

Unbeknownst to me, as I offered my assurances that an asteroid would not strike our planet that day, the media outlets I was speaking to were broadcasting footage of a 10,000-ton meteoroid exploding in the sky above Chelyabinsk, Russia, illuminating the snowy landscape with a flash brighter than the midday sun. The blast emitted a shock wave with the intensity of a nuclear bomb that collapsed buildings and shattered windows. As I prattled on about "how safe we all are and how excited I am about this great day for asteroid science," more than 1,500 Russians were heading to the hospital to treat injuries they sustained in the explosion. In what had to be the most incredible cosmic coincidence in human history, this surprise meteorite collision was completely unrelated to the predicted Duende encounter.

By 9:00 a.m., the Chelyabinsk impact had rocketed up to the top story of the day, the NASA media team was fully up to speed, and we were ready to pivot. I spent the next twelve hours fielding questions, not about Duende and NASA's impeccable predictions, but about Chelyabinsk, and why we never saw it coming.

A few weeks later, I was invited to brief Congress and the White House—suddenly more concerned about asteroids than they had been since Shoemaker–Levy 9 smashed into Jupiter—on Chelyabinsk and the OSIRIS-REx mission. My presentations went well, but the star of the show was undoubtedly the Chelyabinsk sample. Everyone in the room was fascinated with this small chunk of space rock. Its glossy black surface of melted rock impressed upon them the extreme energy unleashed as an asteroid tears through Earth's atmosphere. Seeing their enthusiasm, I left a piece behind, intended to be a gift from the University of Arizona to the President of the United States.

However, White House lawyers decided it was too risky to give him the Russian rock and sent it back. It arrived in Tucson a few weeks later, curiously, at about half the mass of the original sample.

✦ ✦ ✦

Two years after Mike died, I had learned that the job of PI sometimes meant fun stuff—going on television to brag about your scientific accomplishments, for example, or video conferencing with a very excited third grader and his dad about the asteroid they just named. But often, being PI meant difficult decisions and political firefights. In the summer of 2013, we were working furiously to prepare for the mission's confirmation review, which would take place the following May. Even though we had been selected to fly two years earlier, we were only authorized to work up until that milestone. This was NASA's safety valve, a way to ensure they weren't about to throw good money after bad. Technically, I always knew it was possible that our mission could be cancelled—but I was shocked when someone actually tried to make that happen.

To understand the skirmishes that take place in the world of space missions, it's helpful to get a sense of the behind-the-scenes power struggles that occur. Although Goddard

Space Flight Center, the Jet Propulsion Lab, and the Johnson Space Center all operate under the NASA flag, they are constantly competing against one another for mission funding. This environment has led each one to carve out certain niches for themselves to pursue projects and money. For JPL, historically, that had been planetary missions, such as Stardust and Phoenix. Johnson Space Center made its scientific living on astromaterials, curating and analyzing everything from Antarctic meteorites to Apollo lunar samples. For the most part, each organization stayed in its lane. But when Mike—and with him LPL—chose to partner with Goddard, a group that traditionally worked on Earth science and astrophysics missions such as Landsat and the Hubble Space Telescope, leadership at JPL felt spurned. Not only had they been pitching an asteroid mission to NASA for years, but they had long enjoyed the loyalty of the University of Arizona. With Goddard providing mission management on OSIRIS-REx, LPL, they claimed, had switched teams.

Curiously, they seemed to have forgotten that Mike *had* approached JPL leadership on multiple occasions about partnering on the mission. It would have been, after all, the same successful partnership that pulled off the Phoenix Mars Lander. But senior management at JPL declined, choosing to back a group from the University of Arkansas instead. In fact, JPL had jettisoned Lockheed Martin from their asteroid mission, driving them to reach out to UA as a partner. Goddard, eager to pursue this new funding line, welcomed us with open arms.

JPL would get their chance to exact revenge for our perceived betrayal, right before our review. Months before OSIRIS-REx was slated for this important moment, the

Obama Administration announced their intention to send a robotic probe to capture an asteroid, redirect it into a lunar orbit, and send astronauts there to extract a sample. They called it the Asteroid Retrieval Mission, or ARM, complete with a slick computer-generated animation of the concept.

Those of us in the asteroid science community were skeptical that such a mission was even possible. Not surprisingly, models suggested that finding an asteroid with the characteristics that ARM demanded was exceedingly unlikely. The asteroid would need to be about a hundred times smaller than Bennu, which would make its orbit and tumbling motion highly unpredictable. These uncertainties would translate into the need for a time-consuming and expensive test program prior to launch. Still, JPL quickly announced their intention to pursue funding for this project.

What surprised me most as the idea of ARM gained momentum was that no one at NASA had approached OSIRIS-REx for advice on the new program. After all, we were the one team in the agency that had spent the better part of a decade finding and characterizing target asteroids for space missions. It was quiet—too quiet. As the confirmation review approached, friends and colleagues within NASA Headquarters began surreptitiously sounding alarm bells.

"The ninth floor is looking at canceling OSIRIS-REx and replacing it with ARM," more than one person forewarned.

JPL claimed that ARM was a direct response to the presidential initiative and should, therefore, take precedence over OSIRIS-REx. Furthermore, they said they could get it done quicker and cheaper than OSIRIS-REx, while bringing back tons of asteroid compared to our mere ounces. Even I had to admit, the sheer audacity of their claim was impressive. As we readied our response, Mike's absence was thrown into stark relief. These were things that Mike would have handled with aplomb, I often thought. I could see him waving his hand at me. "Don't even worry about this, Dante," he'd have said. "I'll take care of it." And he would have.

We prepared our rebuttal to JPL's outrageous claims, describing in detail our journey over the past decade. After seven years of refining our proposal, we had spent two years finalizing designs, vetting, and developing a meticulous budget and management plan to implement our concept. Our cost fidelity was driven by a realistic assessment of the unknown factors that drove our decisions about the equipment in our spacecraft tool kit.

It wasn't just our earned expertise that put us ahead. Timelines for these missions are dictated by the orbits of both the Earth and the target asteroid, and we were fourteen years ahead of ARM. The statistical likelihood of even finding a suitable target with current telescope assets was minuscule. How could they claim any kind of credible timeline when they didn't even know where they were going?

Our review was chaired by NASA's Deputy Administrator, Robert Lightfoot, which was a good sign for us; cancellation is serious business and it seemed unlikely the deputy would be able to order it. As we presented our case to the review board, which included the director of every NASA center plus a contingent of Headquarters personnel, one issue rose to the top of the list—our laser TAG.

They listed their concerns in a NASA-like staccato of complications. As the chairman read out each reviewer's worry, my brain twitched with fear.

- You are relying too much on Itokawa; you can't be certain Bennu is the same.
- Your laser TAG guidance accuracy is a whopping 10 percent of the asteroid diameter; you haven't considered the asteroid curvature and surface roughness.
- Your laser vendor is untested; they have never delivered flight hardware to NASA, and they are way behind schedule with a mile-long list of technical issues.
- Your simulations are incomplete.
- Your navigators are green.
- *You are going to end up like Hayabusa.*

Those last words echoed through my head, and not for the first time—and certainly not the last—I was sucked into a daydream—a day-*nightmare*, really—about that moment of terror.

In it, everything started out perfect; OSIRIS-REx departed its Safe-Home, fired its engines at CheckPoint and at MatchPoint, and then began its final descent. In a control center somewhere, I received the signal that the spacecraft was heading for the asteroid and then...nothing. Radio silence. And the sick feeling that sixteen years of my life would have been dedicated to a pile of space junk, a memorial to our bravado, one that would endure for millions of years. Sometimes the daydream ended with futuristic visions of school kids on a field trip, maybe five hundred years from now, gawking from a space bus at the mangled wreckage.

Suddenly, Lightfoot's voice forced me back into the moment. "NASA goes into these situations with a belt *and* suspenders on," he said with a slight grin. "The last thing we want is to get caught with our pants down, so to speak, at the moment of asteroid contact."

The direction was easier said than done: *Find a backup solution to laser TAG.*

We accepted their direction and agreed to study options for a secondary navigation system. Our shoulders relaxed incrementally as every other mission element was given hearty approval.

At the end of the meeting, alongside representatives from Goddard and NASA Headquarters, I signed my name on the Decision Memo, a document that gave us the green light to proceed into the next phase. It came with a check for five hundred million dollars. Over the next three and a half years, we would spend that fortune to build and launch the ultimate robotic asteroid explorer.

As I stared down at the memo, studying the three signatures, I recalled the recent challenges we had overcome. Engineering obstacles. Industry subterfuge. The sky literally exploding while millions watched me. And, somehow, we had done it all without Mike.

"Up and out," I thought to myself and to Mike—wherever he was—"after all."

One of the best things about being part of a space mission is being a member of an international community. A year and a half after our successful confirmation, we gathered in the Drake Building to cheer on our friends in Europe as they attempted to drop a lander on comet 67P/Churyumov–Gerasimenko.

Three months earlier, in August 2014, the Rosetta spacecraft reached the comet and went into orbit. The initial images shocked everyone. Just like on OSIRIS-REx, the Rosetta scientists had built a model of their comet, a compilation of all their best knowledge that drove their mission design. They expected the comet's surface to be smooth, its shape defined by gentle hills that gracefully arced from one curve to the next. The reality was a nightmare for lander operations. The surface was highly irregular and extremely harsh, with two distinct lobes separated by a cavernous "neck." It looked nothing like what they were expecting.

I seized on this shocker to wake up my own team. We looked at a slide deck that started with Rosetta's envisioned comet, then the brutal, beautiful beast that was 67P, followed by our best guess for Bennu, and ending with a giant question mark. I looked over at Arlin Bartels, the flight system manager based at Goddard. Arlin had joined the mission at Heather's insistence, after working with her on the Lunar Reconnaissance Orbiter. His passion and dedication for spaceflight was palpable, and he seemed to relish the challenges offered by OSIRIS-REx.

Once our eyes locked, I said, "We have a new risk: Bennu surprises. We need to be ready to be surprised."

In November, Rosetta's probe, called Philae, touched down on the comet, and the deployment looked perfect. The camera captured amazing shots as the lander separated and began its journey down to the surface. But then, unexpectedly, it bounced, landed, bounced again, and landed for a third and final time. This unplanned hopscotching was the result of a double failure—both the "belt and the suspenders," as it were, broke when its anchoring harpoons failed to deploy, and a thruster designed to hold the probe to the surface didn't fire. To make matters worse, the lander's final uncontrolled touchdown left it wedged in a dark crevasse in the comet's ice, just like those death traps that we trained on in Antarctica.

This "nonoptimal location and orientation," in spacecraft speak, meant that the probe would only function for three days—a major loss compared to the original strategy of six months of operations. Despite the challenges, the probe obtained the first views from a comet's surface. It found about ten inches of gravel sitting on top of solid ice. The mass spectrometer detected carbon and hydrogen, sure signs of organic molecules.

Still, it was yet another spacecraft contacting a small body, and another stumble and fall. Drawing on Mike's wisdom, I told everyone Philae's failure was good news for us.

First, the lander bounced! Just weeks ago, the TAGSAM engineers had briefed me on the latest concept for the robotic arm. They had added a spring to the forearm as a way to absorb the momentum of contacting a hard asteroid surface. As a bonus, they said, if the spacecraft reboots at the moment of contact, we will just bounce off the surface, like a kid riding a pogo stick. Philae's roller-coaster ride over the comet surface made this sound like a very good precaution indeed.

Knowing that there were more important lessons we could learn, I called up Ed and told him to pack his bags and head to Europe. "Pick the Rosetta team's brains," I said.

"Learn everything you can about what they would do differently." Constantly seeking new experiences, especially ones grounded in science, he eagerly accepted the assignment. After a whirlwind tour through France and Germany, Ed came back with a clear message: "Add a navigation camera."

This advice aligned perfectly with the backup navigation solution—the one NASA had demanded—coming from the engineers. They called it Natural Feature Tracking. NFT would enable OSIRIS-REx to snap pictures of the surface as it descended to the sample site, compare them to an onboard catalog that we preloaded in its memory, and determine its location relative to the planned trajectory. After analyzing a couple dozen images, the spacecraft would then accurately update the CheckPoint and MatchPoint maneuvers to guide itself to the sample site.

This extra "NavCam" was exactly what we were looking for. It felt like adding the slick GPS package to a new car.

But almost immediately, Lockheed management came back to say they simply didn't have the programmers available to write this new software. Since NFT was a backup technique only, they reasoned, we could build and install the new navigation cameras and create "stubs" in the guidance software, where a full NFT code could be installed in-flight if needed. Like a standing army waiting in reserve, the programmers would be available later if they were needed, they assured me. Something we wouldn't be sure of until we arrived at Bennu in four years.

It turns out, building a spacecraft is a lot like building a high-performance race car. Both have to work in extreme environments with no room for failure. Success often comes down to millisecond precision. And as soon as one problem is solved, another emerges that sets us scrambling for solutions once again.

We had solved one problem after another, especially in the sample collection systems. We had evolved from "point and shoot" to "laser TAG with an NFT chaser." We had built a reverse-vacuum-cleaner-pogo-stick robotic arm to hoover up sample from the surface. Of course, it was time for the last piece of the puzzle, getting the sample across the finish line back on Earth, to throw us for a loop.

Up to that point, Lockheed had a mixed record of delivering their sample-return capsules safely back to Earth. Everyone was familiar with the failure of the Genesis mission; its capsule crashing into the Utah desert was a constant feature of my nightmares. We all knew this was the result of human error—the gravity switch that released the parachute had been installed backward, rendering it useless.

Their success—the capsule on Stardust, which landed perfectly—informed our sample-return capsule. We incorporated a key lesson from Genesis as well: Test the entire system to make sure all components are installed correctly and fully functional. And in fact, these tests quickly revealed major flaws.

The first sign of trouble showed up during the "drop test." For this rudimentary but effective test, we simply dropped the capsule out of a helicopter to verify that the parachutes deployed. This occasion was exciting, not just for the team, but for our growing legion of followers around the world. We decided to film it and use the footage for some cool social media posts, giving the world a preview of the events to unfold nine years in the future, when OSIRIS-REx would deliver samples of Bennu back to Earth.

We hired Arizona Public Media to provide video coverage of the event. The test was remote, and we gathered in a conference room at Lockheed to watch. I knew the timeline, down to the second, of when the parachutes should release. With all eyes on the screen, I watched the capsule drop out of the helicopter and started counting down to chute release. As the seconds ticked off, I felt my heart rate tick up.

If I am this nervous today, I thought, *what I am going to be like during the real thing?*

Finally, the moment of chute release approached, and then—it passed. No parachute. My eyes shot to Arlin. That day he appeared a bit disheveled, with dark circles under his eyes and a scruffy beard on his face. I could tell he had spent long nights working through difficult engineering challenges of getting this test to work. His eyes focused on the screen; he felt my gaze.

"This isn't like a rocket launch. It's subject to some random forces," Arlin assured me. Then, after what felt like an eternity but was, in reality, 3.2 seconds later, the little drogue chute released, followed almost immediately by the main chute. As a result of the delay, the deployment occurred at an elevation a couple hundred feet lower than intended. This meant that the video footage was spectacular, as the camera crew was able to zoom in on the chute deployment at much higher resolution than expected. Unfortunately, since this footage recorded an anomaly, NASA would not want to release it to the public. Instead, we shipped it over to the engineers to help them solve the problem.

Their first notion was that the culprit was the tape wrapped around the bridle wires. So, back to the helicopter we went, after we revamped the tape configuration to fix this problem. Same test, same result.

The next suspect turned out to be the right one. A buckle that constrained the bridle wires was interfering with their release. The engineers needed to fabricate a new design and reconfigure the parachute. Two months later, we were back to the helicopter. This time, it worked. The parachute released on time, as expected, with no sign of snagged cords.

As we wrangled with an unruly drogue parachute, the gravity switches that release the parachutes also started to act up. These "g-switches" sense when the capsule has reached a critical deceleration and then they release the chutes — if they are installed properly, that is. Once the capsule was fully assembled, we transferred it to the University of Colorado for centrifuge testing. A centrifuge is a device that rapidly spins a package at the end of a long arm, to increase the acceleration and simulate high-gravity environments for extended periods of time. Astronauts have to spend time on a centrifuge to prove they can withstand the intense accelerations and remain functional during rocket launches and capsule reentry. We were going to treat our capsule the same way, for the same reasons.

We loaded the capsule into the cradle at the end of the spinning arm and started to bring it up to speed. We reached the critical acceleration when the g-switch should deploy and — nothing. The switches remained closed. The sinking feeling in the pit of my stomach returned.

I turned to Arlin and asked him, more sarcastically than I should have, "You sure you installed them in the right direction?"

He glared at me, then promised to get to the bottom of the problem. The first solution was to completely disassemble the unit and repair and replace all g-switches. With only seventeen months to launch, I cringed at this step backward, watching painfully as the capsule was broken down into its constituents, which were laid out like an autopsy patient's innards. Such issues are why we have schedule margin, I reminded myself.

After an interminable seven months, the g-switches had been removed and replaced with fully checked parts. Post-installation tests looked good. We started the final closeout and shipped the rebuilt capsule back for retest.

The result was concerning, to say the least. When the capsule reached the appropriate acceleration, the switches stuck. The new suspect was debris inside the switch chamber. These switches are like little balls on a spring. Once the force is strong enough, the ball compresses the spring, flipping the switch open. The engineers posited that some dirt was interfering with the spring, making it stutter before building up enough force to push past the frictional barrier.

As I listened to the detailed report, my heart stuttered along with the switch data. I envisioned myself riding that capsule back down to the surface of the Earth, waiting with bated breath for the parachutes, only to realize that the stuttering vibration was preventing their release. I scanned the room and noted all the determined faces. We were going to get to the bottom of this.

The engineers started casting about for an alternate supplier of g-switches, while simultaneously beginning to interrogate the original ones. They scanned them with X-rays and cut them open to examine them under a microscope. They were squeaky clean. Lockheed and Goddard called all hands on deck. Suddenly, it seemed quite plausible that the moment of terror wouldn't be TAG after all, but landing—or crashing—in the desert.

With five months to launch, I got a call from Arlin. "We have figured it out!" he exclaimed. "It wasn't the switch at all; it was the way we had designed the centrifuge test."

"What do you mean?" I queried. "Is there a problem with the acceleration?"

"No," he replied. "The test configuration created a side load on the g-switches, pushing the bearing against the wall, causing the observed chatter. The solution was to restructure the test to simulate the flight-like acceleration profile more accurately."

With the unit reassembled and the new test in place, we crossed our fingers and ran one more test. Everything worked perfectly.

✦ ✦ ✦

In May 2016, it was time for OSIRIS-REx to head to Kennedy Space Center. It was time, in fact, for all of us to make our way there, to spend the next four months preparing for launch.

To mitigate any contamination, the shipping container that would transport the spacecraft from Lockheed Martin to Buckley Air Force Base was flooded with a near-continuous purge of high-purity nitrogen gas. With the cargo loaded and swathed in its protective gas bubble, we were joined by the Douglas County Sheriff's Office, who escorted us to the base, where an Air Force C-17 Globemaster III aircraft awaited us on the runway. We felt like royalty with our full entourage of security personnel. As Heather and I rode along in a van behind the eighteen-wheeler hauling our spacecraft, I glanced at the Rocky Mountains on the horizon and then looked over at her.

"We actually did it," I marveled. "We built the world's greatest asteroid explorer."

"We sure did," she confirmed, "and now we get to send it on its expedition into the great unknown."

Once we arrived at Buckley, the shipping container was removed from the trailer, placed on the tarmac behind the aircraft using a forklift, and carefully pulled aboard with a winch. The Air Force pilots invited me up into the cockpit to swap war stories. I told them about the asteroid adventure ahead of us, then listened with awe as they described their last mission, delivering supplies to US troops fighting in Afghanistan. They were thrilled at the historical significance of touching down on the Kennedy Space Center Shuttle Landing Facility. They were even happier that no foreign adversary was shooting at us as we came in.

Offloading from the C-17 was essentially the reverse of the loading procedure. The spacecraft was taken to the Payload Hazardous Serving Facility, our home for the next few months. Here we performed final checkouts and squeezed in last-minute installations. Two copies of the lasers for our game of TAG showed up just in time and were the last components to get bolted on. After fueling the tank with 1200 kilograms of hydrazine, we encapsulated OSIRIS-REx in the payload fairing, the nose cone that protects the spacecraft from exposure to the atmosphere.

Three months later, OSIRIS-REx was perched atop the mighty Atlas V rocket, ready and willing to leave this world behind on its hunt for asteroid Bennu.

Dante with OSIRIS-REx and the Atlas V fairing at Kennedy Space Center in 2016. Credit: Dante Lauretta personal photo.

CHAPTER 9

GO OSIRIS-REX

On the morning of launch day, I had fifteen minutes to myself. That's how long it took to drive from Cocoa Beach, where we were staying with our families, to the Cape Canaveral Air Force Station. I had carefully assembled a playlist for this day, despite knowing there would be scant time to enjoy it. It began—obviously—with Rush's "Countdown." With the midmorning sun overhead and the Atlantic Ocean glittering beside me, my pulse sped up to keep time with the song's synthesizers. Geddy Lee sang, "Lit up with anticipation / we arrive at the launching site."

As I drove past the now-familiar Florida beaches, I thought of Mike. His fondness for the beach was well-known, and we'd had a fun-filled week together at a science conference in Rio de Janeiro. He would have loved to be here today.

To get to the rocket control tower, I passed through a security checkpoint, where a man with a machine gun glanced at my ID and waved me in. I smiled as I pulled past the entrance sign, which today read, "Go OSIRIS-REx!" The mission logo filled another nearby billboard. Just ahead, I could see the nose of the Atlas V rocket pointing up to space like a baseball player calling a home run. I imagined OSIRIS-REx tucked inside, waiting anxiously to fulfill its destiny.

I pulled into the parking spot labeled Principal Investigator, and snapped a photo of the plaque for posterity. The United Launch Alliance, the private launch company that would send OSIRIS-REx into space, had been nothing but hospitable. They certainly treated me like someone who paid 183 million dollars for one of their rockets, anyway. During our four-month sojourn on the Space Coast, residents had been even more welcoming, approaching us in restaurants and grocery stores, impressive in their depth of knowledge and enthusiasm about the mission. A stylist at a barbershop in Cocoa Beach recognized me immediately, peppering me with nonstop questions about Bennu, the spacecraft, and the launch during my trim. After that, I started carrying commemorative pins to hand out.

But for all the support and fanfare, the last few weeks had been full of reminders that missions like ours can and do fail, and often at the very last, most heartbreaking minute—on the launchpad.

Six days before launch, I was in a conference room at the Neil Armstrong Operations and Checkout Building preparing for our flight readiness review, the meeting for launch teams to discuss last-minute concerns. Sitting there, my attention was drawn to the emblems that adorned the walls, logos of the hundreds of spacecraft that NASA had launched during the past half century. A few of the logos were printed in black and white, not in vibrant color like the others, and I tried to figure out what they might have in common. When I saw the monochrome logo of Glory—a 2011 Earth-observing satellite whose rocket malfunctioned minutes into flight and plummeted into the Pacific Ocean—it hit me: the grayed-out logos commemorated failures. How would OSIRIS-REx be memorialized? I wondered. In celebratory color, or solemn black and white?

As if the universe were reading my mind, the table began to tremble. A low rumble growled beneath our feet. The roar grew louder, culminating in the stomach-turning sound of an explosion, a noise that's terrifying in any circumstance, but even more so when your spacecraft is next door. Almost every cell phone in the room buzzed against the table. I hesitated and then turned mine over. "SpaceX just blew up on the pad!" the text message read.

Slowly, my colleagues and I pieced together what had happened: A mile east of the launchpad, where members of our team were putting the finishing touches on OSIRIS-REx, the SpaceX Falcon 9 had blown up during a routine test. The rocket collapsed in on itself, igniting a massive fire. Then, the two-hundred-million-dollar Facebook communications satellite it was carrying hit the ground with its own load of fuel, setting off a series of secondary blasts. From the parking lot, I watched billows of black smoke drift toward our pad, pushed by the relentless sea wind. Word spread that no one was hurt, and Air Force support were on their way to make sure OSIRIS-REx was safe. People began to head inside and resume the meeting, like so many office workers returning to their day after a fire drill. I followed, doing my best to ignore my nerves, crackling like live wires.

The flight readiness review reconvened and concluded with a unanimous "Go" decision, an occasion that shouldn't have been anything but triumphant. Instead, I left worried and grim.

A week later, I stood in awe as the twenty-one-story-tall Atlas V 411 rocket emerged from the cocoon of the vertical integration facility and rolled past me on its way to the launchpad. It was a strange-looking launch vehicle, with only one booster strapped to its side. I'd spent lots of time in the past few days explaining its configuration to reporters, who saw it and asked how it wouldn't spin out of control. (In a nutshell: The main engine tilts inward to redirect its own thrust and allow Atlas to rise perfectly straight into the sky.) Before I headed inside to the control center, I gave a nod to the words painted on the side of the rocket: "Colleague, Friend, Visionary," in honor of Mike.

The launch window would open at precisely 7:05 p.m. when the path between Earth and Bennu aligned with central Florida. After it left the pad, Atlas V would accelerate to 25,000 miles per hour, defying the gravitational pull of the Earth. Fifty-five minutes after launch, the rocket would release OSIRIS-REx, placing the spacecraft into orbit around the sun. Parts of the rocket would fall back to the Earth into the depths of the Atlantic Ocean, while the rest would remain floating through space. OSIRIS-REx would then travel around our solar system for a year, so it could return to our planet and use Earth's gravity field to propel it one billion miles to Bennu.

I took a seat at my designated console. Before me were dozens of windows and menus and widgets, all obsessively keeping tabs on the myriad of rocket subsystems necessary to launch into space—aka nerd heaven. Kate, the kids, and the rest of our families were viewing the launch from the rooftop several floors above me in the Atlas Spacecraft Operations Center. As much as I'd have loved to watch their eyes as the spacecraft ascended into the sky, I was also grateful for the relative calm of the control center and this comfortable chair.

There had been moments of reprieve in the last few weeks—cookouts with my colleagues, playing on the beach with my boys—but mostly it had been a blur of overstimulation, like a wedding where I had no control of the guest list. There was even a signature cocktail, the Blue Bennu—tequila, blue curaçao, and fruit juice—served at the bar at the Hilton Hotel where many of our guests were staying, and party favors: commemorative coins, stickers, patches, posters, and pins.

My mom and her husband had flown down from Arizona. My brothers were here with their wives and children, and my sister-in-law brought her fifth grade science class on an unprecedented field trip to the Kennedy Space Center. Just like with my actual wedding, I felt torn in a million different directions, worried I wasn't enjoying the moment enough.

I also had work to do—lots of it. Lectures. Interviews. Appearances. I'd learned that being the principal investigator of a mission about to launch was a lot like being its principal cheerleader. For the most part, I enjoyed the opportunity to

Dante and Heather at the launch site. Credit: Dante Lauretta personal photo.

brag about our team and talk shop about OSIRIS-REx, but I couldn't deny how exhausted I was. The day before launch, on my way to a University of Arizona alumni event, I had walked straight into a glass door. Luckily, no one saw and the only thing I had injured was my pride.

Now, the only thing left to do was to send our spacecraft on its way. I put on my headset, which was filled with the voices of the launch team, stationed in front of screens around the world, as well as the public audio feed. It sounded like a dinner party; dozens of conversations all happening at once. I rolled up my sleeves and smoothed my tie, white with red-and-blue OSIRIS-REx logos checkered across it.

Dante with the Atlas V rocket that launched OSIRIS-REx.
Credit: United Launch Alliance.

At first, I kept my eyes trained on the cryogenic liquid oxygen levels, the culprit in last week's SpaceX explosion. Just as my eyes settled on the gauge, a major problem emerged near the rocket's second-stage oxidizer tank, which was actively being filled with highly explosive liquid oxygen rocket fuel. I tried to shake last week's image of black smoke out of my mind. At 6:00 p.m., with around an hour remaining until the opening of the launch window, the liquid oxygen tank reached flight level and I felt a pound of worry slide off my shoulders.

At 6:35 p.m., the 45th Weather Squadron presented their final briefing, indicating a 90 percent chance of acceptable weather. Weather readiness was a "go."

At 6:55 p.m., the spacecraft transitioned to internal power, running on its own batteries for the first time as the "umbilical cord" to ground power was cut. I pressed a fist to my lips as the final "go, no-go" poll unfolded in my headphones. It was a string of "go" across the board.

At 7:01 p.m., the terminal countdown began at T minus four minutes. On the launchpad, Atlas V's propellent tanks were pressurized and the flight termination system, tasked with destroying the vehicle in case of a potentially explosive emergency, was armed.

"Range Green," came the call at T minus one minute, meaning everything was still a "go."

"Launch director—" said the lead system engineer.

"You have permission to launch."

With launch less than one minute away, I became perfectly still. I silently thanked the thousands of people who had worked together over the last twelve years to design this mission and this spacecraft, the tens of thousands of people gathered today in Florida, and the hundreds of thousands of people who submitted their names to be etched onto OSIRIS-REx, joining us in spirit on our journey to Bennu.

"T minus twenty-five. Status check—"

One more chance to call it off.

"Go Atlas." The first stage was ready to launch.

"Go Centaur." The second stage was prepared to take over when called upon.

And finally, the words I had waited a decade to hear:

"Go OSIRIS-REx."

And then it happened, just like in the movies, just like in the Rush song.

"T minus ten. Nine. Eight. Seven. Six. Five. Four. Three—"

At T minus 2.7 seconds, the Atlas V's main engine roared to life, generating 860,000 pounds of thrust from its twin nozzles. I held my breath. I closed my eyes. I felt my consciousness expand beyond the room, beyond the facility, beyond the confines of place and time. I felt the waves of billowing smoke, the heat of the fire erupting from the engines, the launch vehicle gently, perfectly lifting from the pad. I felt my body rising alongside OSIRIS-REx, soaring into space together on our cosmic expedition. I did not hear the final seconds tick away.

"Liftoff of OSIRIS-REx," the public audio feed announced. "Its seven-year mission: to boldly go to the asteroid Bennu and back."

The whole thing was flawless. Beautiful. The smoke produced by burning liquid oxygen and kerosene built a tower to the sky. The rocket disappeared into the heavens.

"Fare thee well, my friend," I whispered.

OSIRIS-REx was on its way.

PART III

CHAPTER 10

ARRIVAL

TUCKED INSIDE THE NOSE CONE of the Atlas V rocket, OSIRIS-REx was ready for its journey. The ignition of the rocket engines caused raucous vibration and deafening noise, and the spacecraft groaned under the pressure of the accelerating liftoff, which exerted six times the force felt on Earth. Within minutes, the first stage of the rocket detached and plummeted into the Atlantic Ocean, followed by the fairing. Then, the pressure relented, leaving the spacecraft weightless in orbit.

From OSIRIS-REx's view in space, the Earth appeared as a beautiful blue-and-white sphere with swirling clouds and a visible atmosphere. As it passed over Africa and into the night, the bright lights of cities and towns became visible against the blackness of the cosmos, the stars more prominent.

As the Australian continent came into view, the second stage fired its engine then released OSIRIS-REx into the solar system, racing away from the Earth at 12,000 miles per hour. Explosive bolts discharged, and mechanical latches released the solar arrays and angled them toward the Sun. The spacecraft's solar panels drank up the sunlight, replenishing its batteries. The flow of electricity needed to power the other components began. The vastness of space beckoned with the promise of adventure as the spacecraft set off to explore the unknown. Now, all we could do was wait.

✦ ✦ ✦

After one year in orbit around the Sun, OSIRIS-REx detected our bright planet in the distance. Earth grew in size and became a well-resolved crescent, its limb softly backlit by the Sun. During its approach to Earth, dawn broke, and the stark white landscape of Antarctica streaked below, inky blue oceans blooming from its jagged shores. The force of the descent into Earth's gravity well was so enormous, the spacecraft shuddered under the pressure.

That same morning, I pulled into the parking lot at the Drake Building. On my way into the office, I made sure to note the digital countdown clock in the lobby:

TIME UNTIL EARTH GRAVITY ASSIST
1 hour, 49 minutes

I paused to enjoy the thrill those words sent down my spine.

On the day of our launch, it had read *378 days, 15 hours, 55 minutes*. Time had flown by—and now our spacecraft was about to as well. On that morning, at 9:52 a.m. Tucson time, OSIRIS-REx was going to zip past Antarctica, flying 10,711 miles over the southernmost continent, just south of Cape Horn, Chile, before zooming north out over the Pacific Ocean. Like so many planetary explorers before us, we were using Earth's gravity to slingshot ourselves on a path toward Bennu. We called this maneuver EGA, short for Earth gravity assist. The EGA was critical to rendezvousing with Bennu the following year.

With a healthy spacecraft and the crew in high spirits, we were ready for the next major event on our journey out to Bennu. Although the Atlas V rocket provided all the momentum required to propel us to Bennu, OSIRIS-REx wasn't in the right orbital plane. Bennu's orbit around the Sun is tilted at a small angle from that of the Earth. So, we needed an extra boost from our planet's gravity to bend its path and catch the asteroid. Gravity assists are like speed-up boosts in video games: Hit the right spot in space and your spacecraft goes zooming off in a new direction. As a result of the flyby, the velocity of the spacecraft would increase by a whopping 8,451 miles per hour. To get this same boost using the spacecraft's rocket engines would have required more than twice the total amount of fuel we were carrying in our tank.

In addition to placing OSIRIS-REx into interplanetary overdrive, the EGA was the first opportunity to test-drive its science instruments. To ensure that the approach path through Earth's gravity well—the depression in spacetime due to the strong gravitational force of our massive planet—went exactly as planned, the spacecraft waited approximately four hours to get to the sunlit side of the planet.

With the opportunity to open its eyes, the spacecraft

turned its instrument deck back toward home again and initiated imaging and spectral mapping of the Earth. The data revealed endless seas, a swirling atmosphere full of cyclonic storms and a panoply of gases—methane, oxygen, carbon dioxide, and ozone. The land masses were inlaid with minerals and a photosynthetic compound that harnessed the energy of the Sun, sustaining a rich biosphere.

Everyone was excited for the images of our home world and the spectroscopists were anxious to point their instruments at the Earth and its atmosphere full of strong spectral fingerprints. For the first time since launch, the Drake Building crackled to life. Science team members from across the globe gathered to participate in this historic mission milestone and practice performing the daily activities that would be necessary during the asteroid encounter. For months, I had been looking forward to the Earth-observation campaign as a way to cultivate camaraderie.

Estelle Church, our spacecraft systems engineer, was visiting from Lockheed Martin. Estelle had blond hair and always wore spangly, eye-catching jewelry that matched her energetic personality. Lively, loud, and full of verve, she was a highly skilled and time-tested engineer, with a passion for designing, building, and operating our spacecraft. She was friendly and approachable, the perfect liaison between the engineers and scientists in the room.

Everyone was camped out in our science operations area, a glorified conference room with cubicles lining the walls. (This space was also affectionately known as the "doughnut-distribution center," where we would gather every Friday morning to discuss the latest mission status and partake in our weekly allowance of sugar and carbs.) Estelle's computer displayed a dashboard, which was constantly updated with the latest data from the spacecraft. I peered over her shoulder and scanned the temperatures, currents, voltages, instrument states—down the list, everything looked perfect.

Just after 6:00 p.m. that evening, a small group of us sat around the conference table in the doughnut distribution center, the anxiety palpable as we awaited the first views of Earth. We weren't nervous; we were anxious. I sat in my chair crossing and uncrossing my legs. I had trouble focusing on anything else—my thoughts were consumed by the upcoming event, and I constantly flicked my gaze over to Estelle's screen.

"How's it going?" I asked, unable to stop myself.

"We're on twenty-five right now," Estelle replied, referring to the receiving dish we were using to listen to the signals from OSIRIS-REx. This 110-foot-wide satellite dish was part of NASA's Deep Space Network, an international array of antennas that supports interplanetary spacecraft missions along with radio astronomy. It consists of three deep-space communications facilities placed approximately 120 degrees apart around the world: the Goldstone Complex outside of Barstow, California; the Madrid Complex west of Madrid, Spain; and the Canberra Complex southwest of Canberra, Australia. Once you get far enough away from the Earth, no matter where you are in the solar system, you can always see one of these stations.

"California?" I asked, knowing full well that we were using the Goldstone Complex.

"Goldstone," she confirmed, both of us talking simply to break the tense silence. "Any minute now."

As the interminable minutes ticked by, my mind cycled through a list of concerns. Our camera suite was specifically

Dante at the Goldstone Complex of the Deep Space Network.
Credit: Dante Lauretta personal photo.

designed for Bennu, which is one of the darkest objects in the solar system and has a reflectance similar to that of highway asphalt. Given its sensitivity, the bright continents and billowing white clouds of the Earth could potentially saturate our detectors, resulting in an overexposed mess. This would not be a problem from a scientific perspective, as the cameras were not built to study the Earth. Nevertheless, this first image would be crucial in demonstrating our capabilities and I was hoping for an iconic representation of our home world. I also knew, in a visceral way, that it would provide an unparalleled morale boost for the team, one definitive piece of evidence that we were up to the challenge ahead of us.

I perched over Estelle's left shoulder, scanning the engineering data that continued to stream across the display. The data should be down already, why weren't they on the screen?

"This is the longest five minutes of my life," she confessed, echoing the sentiment of everyone in the room.

After thirty minutes of waiting, we knew the second the Earth appeared on Estelle's computer because she yelped: "Oh my god! Wow!"

In Arizona, as the data came down, everybody clamored to Estelle's side of the table. Gazing at our gorgeous planet, we let out a collective gasp. There it was, planet Earth, beaming back at us from outer space.

The Earth as seen by OSIRIS-REx. Credit: NASA/Goddard/ Lockheed Martin/University of Arizona.

Staring at the screen, I let myself imagine I was *inside* OSIRIS-REx, behind its cameras, gazing in wonder at our home world. There, encapsulated in those few thousand pixels, was us—all eight billion of us—along with the myriad of amazing life-forms with which we share our planet. It looked so fragile, so delicate, so beautiful. For a moment, I considered what might change if everyone had a chance to see the world as I was seeing it at that moment.

Spontaneous applause snapped me back to the present moment. Tears streaked down some faces, and I could tell that a few of my friends had also been imagining themselves in space. Together we marveled at our collective accomplishment, amazed that everything had gone as planned.

Now, the instruments were working at full speed, and for the next three days we interrogated the Earth and our Moon, putting the hardware, our ground system, and our team through their paces, like a scrimmage game before the big tournament, building skill and confidence for the upcoming encounter with Bennu.

Within days, the splendor, the sharp glare of the Earth was gone, and OSIRIS-REx headed off into deep space again, this time on a path to catch its target. Like the whiplash turn of a roller coaster, the trajectory of OSIRIS-REx had been bent upward by six degrees, now matching the orbital tilt of Bennu, and it swooped away from Earth. The hunt had begun.

After the excitement of EGA, the clock in the Drake Building's lobby was reset:

TIME UNTIL BENNU ARRIVAL
438 Days

✦ ✦ ✦

In the years since the first Hayabusa spacecraft returned its asteroid dust, JAXA had revamped and launched a sequel to their ill-fated but ultimately successful mission. Over the years, the Hayabusa2 and OSIRIS-REx teams had developed a strong working relationship. We all knew that our cooperation increased the chances of success for both missions.

In August 2018, I was in Japan for the Hayabusa2 landing site selection meeting—there to help but also to glean insight about our own site selection. Hayabusa2's target was asteroid 1999 JU3, now named Ryugu after the dragon palace from Japanese folklore, a dark, near-Earth asteroid in a similar orbit to Bennu. In taking part, I was now helping explore the two largest carbonaceous near-Earth asteroids simultaneously.

Prior to launch, JAXA had much less information about their asteroid than we did. Most critically, there were no radar data from which to render the size and shape. It was as if Ryugu occupied that uncharted part of the map that proclaimed, "Here There Be Dragons." In all the astronomical observations, Ryugu only ever appeared as a point of light. Hayabusa2 based their mission concept on the brightness variation as the asteroid rotated and went through different illumination angles.

In late July, right before I arrived in Japan, JAXA had obtained the first close-up view of Ryugu. As I studied those images, I was struck most by Ryugu's shape. It had that recognizable spinning-top form and appeared remarkably similar to what we expected from Bennu, even though it was more than twice as wide.

"That shape looks familiar," I said knowingly as I leaned over to Hayabusa2's lead scientist, an old friend.

"Yeah," he replied, his face breaking out into a wide grin. "At first, I thought we made a wrong turn and ended up at Bennu!"

Prior to Hayabusa2, we had never seen one of these objects up close. We had designed the entire mission profile around the spinning-top shape, so to see it in Ryugu reassured me we were on the right track. This result confirmed all of our suspicions—that Ryugu, and most likely Bennu, were really spinning piles of rubble. Like coarse droplets in the cosmos, these asteroids were nothing more than loose accumulations of space debris. Splashes left in a cosmic pond when two much larger asteroids crashed into each other, hundreds of millions of years ago, out in the main asteroid belt. Millions of objects like our two targets coalesced as the impact debris slowed down and ultimately collapsed back into these clumps. Gravity always wins.

One aspect of Ryugu that defied expectations, however, was its rough-looking surface. The asteroid was covered in boulders of all size scales, from over three hundred feet across down to about a foot and a half, the current imaging-resolution limit. The boulder density was about twice that observed on Itokawa. Even though we were here to pick several landing sites for Hayabusa2's armada of rovers, landers, impact experiments, and sample collectors, the current resolution was not high enough to perform the needed safety assessment. They were struggling to identify pixel-sized boulders by applying slick image processing analysis, really pushing beyond what I felt was credible. A sense of dread permeated the room and my psyche; I hoped we would not have to deal with these challenges.

The downside of being in Japan was that I missed being with my colleagues back in Tucson as *our* first images of Bennu were downlinked. That night at 2:00 a.m., I was snuggled in my tiny hotel room in Sagamihara when they patched me into the auditorium back at the Drake Building. They heard me whoop and holler, probably waking up my hotel neighbors, as the target I had dreamed about for the past fourteen years became real. Almost two years after launch, we saw our asteroid.

The spacecraft had once again opened its eyes, scanning the stars with its long-range camera. PolyCam took the first shot from a distance of approximately 1.4 million miles. After two years in space, Bennu was now a single pixel moving across the field of stars.

The single dot was unmistakable to those of us who had trained our eyes where to look. Its presence meant that we had done it—we had made it to Bennu. Over the next few months, Bennu transformed from a mere pixel into a full-fledged world. It was a rocky, rugged planet with a rough surface and a spherical body. Driven by a spinning motion that propelled loose material to its waistline, a distinct bulge wrapped around its equator.

Now that we had come into instrument range, it was time to put our science payload to work. We implemented a whole set of observations of Bennu and its surroundings throughout our approach. The most important of these was the Natural Satellite Search campaign—our investigation to ensure the space around Bennu was safe for OSIRIS-REx to roam. This operation enlisted both PolyCam and Map-

Cam. We also added the NavCam, which was slowly but surely getting pressed into science duty for good measure.

As the surface came into clearer view, a few pixels blinded the cameras, brighter than anything we had imagined when we designed it. And yet, other areas were darker than any material ever documented in the solar system. These black boulders hung off the limb of Bennu, barely clinging to its surface as the asteroid spun faster and faster, compelled by the relentless sunlight that imparted a small but constant torque to its body.

With Carl in charge, we surveyed Bennu's immediate environment for any signs of dust, a natural satellite, or unexpected asteroid hazards that would have consequences for spacecraft safety. Carl's report came back with the definitive conclusion: no evidence of hazards anywhere near Bennu. The environment looked pristine. We were clear to proceed with arrival.

As we inched ever closer to the asteroid, the science data poured in. We measured Bennu's reflectance, its light curve (how its brightness changes as it rotates), and its phase function (how its brightness changes as it goes through phases like the Moon). Everything looked as expected.

Throughout the approach phase, the Drake Building was a hive of activity. Science team members rented apartments, brought their campers, or bunked with friends so they could be there around the clock. No one wanted to be left out when the first big discovery came in. We systematically ran through every subsystem and checkout of the spacecraft's two LIDAR systems, the lasers for our game of TAG. They appeared fully functional and ready to play.

The cameras weren't the only instruments scrutinizing

Bennu. Starting shortly after the first PolyCam detection, our spectrometers caught the asteroid in their field of view. One busy afternoon, the scientists in charge of analyzing data from the two spectrometers came rushing into my office, total glee slipping through their poker faces. They slapped a piece of paper on my desk showing a graph with some squiggly lines on it. What those lines conveyed was thrilling. Starting out as a straight line, the graph plunged down in a dramatic fashion, indicating a decrease in the strength of the signal. That drop was a spectral signature, a clear indication that a molecule on Bennu was absorbing some of the sunlight hitting its surface. This spectral band was centered on a wavelength of 2.7 microns, deep into the infrared, beyond what human eyes can see. This wavelength range corresponds to the energy in the bond between hydrogen and oxygen, meaning water was present on the surface of Bennu! Instantly, our hypothesis that asteroids like Bennu may have delivered the water in our oceans gained strength.

"What about the mid-infrared?" I asked, referring to the companion instrument that looked at heat radiating from the asteroid.

"Great news there as well," they replied. "The mid-infrared spectra show strong evidence for clay minerals."

This *was* great news. These compounds incorporate water directly into their crystal structure. Early in solar system history, this water existed as ice, frozen as snowflakes and icy crusts in the protoplanetary disk. Ice is not stable in the inner solar system, where Earth resides, unless the body is wrapped in a protective atmosphere. In order to get water to Earth, it needed to be locked up in a mineral that could hold on to it at the searing temperatures of Bennu's surface.

Clay minerals bind the water, allowing Bennu to retain it at temperatures well above those where ice or liquid water would be stable.

"Scientific pay dirt," I stated. "It looks like Bennu is composed of exactly the material we set out to collect."

Later that night, alone on my back patio, I stared up at the stars as the importance of that day's discovery sank in. When I wrote the closing arguments for our proposal to NASA eight years earlier, I had promised that we would provide "unprecedented knowledge about presolar history through the initial stages of planet formation to the origin of life." Bennu's water-soaked surface made me think that we might actually be able to deliver on that promise.

✦ ✦ ✦

For two months, OSIRIS-REx had been hurtling toward Bennu at a dizzying speed of 1,100 miles per hour. If it didn't slow down, it would shoot right past its target. In preparation for its first deceleration maneuver, the spacecraft ignited its main engines, superheating a layer of catalytic metal.

Once the engines reached their peak temperature— somewhere in the thousands of degrees—the spacecraft opened its control valves and sent hydrazine rocket fuel flowing through the pipes that led to the main engines. When the fuel came into contact with the hot metal, it exploded immediately. High-pressure gas ripped through the four engines' throats and burst from their nozzles.

For eleven and a half minutes, the spacecraft rattled as five hundred pounds of fuel released its chemical energy, slowing it down by 785 miles per hour. Over the next six weeks, OSIRIS-REx fired its engines three more times, decreasing its speed just a little bit more each time. Within Bennu's microgravity field, even the slightest forces could deflect the spacecraft from its target. But OSIRIS-REx stayed true.

Every day, we got a little closer to Bennu. In late October, we reached an imaging milestone. We used something called a super-resolution algorithm to combine eight images and produce our highest-resolution view of the asteroid yet. In these shots, Bennu stretched one hundred pixels across.

Sitting in the near-empty Drake auditorium, I stared at

Bennu as seen on arrival. Credit: NASA/Goddard/University of Arizona.

one of these pictures on my laptop screen in stony silence. Even at this crude resolution, I could tell Bennu was more like Ryugu than I had hoped. Its surface was clearly very rocky and rough, with a landscape dominated by boulder-strewn regions. I couldn't spot a single smooth section anywhere.

"Crap," I muttered.

Estelle's head snapped up from her laptop, where she had been working quietly in a back corner. "What's the matter?" she asked.

"You don't want to know," I replied quietly.

Soon, Estelle was sitting in the seat next to me. I zoomed in until Bennu filled the entire screen. Stabbing a finger at the asteroid, I said, "*That* is what's the matter."

Estelle squinted at the photo. "But where's the beach? You promised us a beach—so where is it?"

"Maybe it's on the other side of the asteroid," I said with a deep sigh. "Let's get the rest of the data down so we can find out."

The rest of the data *did* contain some reason for optimism, namely that we had nailed the asteroid's shape, diameter, rotation rate, and pole positions. But there were some surprises too. The first had to do with the size of the prominent boulder near Bennu's south pole, the only one that was apparent in the ground-based radar data, the only one of that size that we had surmised was present on the surface.

We had named this feature Benben in honor of the rock that emerged from the primordial water in Egyptian mythology, the perch for the Bennu bird at the dawn of creation. Based on the radar data, we estimated it to be about thirty-three feet tall. The new spacecraft data showed that the boulder was closer to 130 feet in height. Not only that, but as the imaging resolution increased with each passing day, we began to see that Bennu was covered in giant boulders the size of our original estimate of Benben—2,750 of them to be exact—with nary a beach in sight. It was becoming abundantly clear that a wide-open, hazard-free landing strip did not exist on the asteroid and delivering the spacecraft safely to the surface was going to be much more complicated than we had anticipated.

✦ ✦ ✦

After a two-year pursuit, the hunter reached its quarry. The spacecraft's previous maneuvers had been a display of brute force, consuming hundreds of pounds of rocket fuel. But for the final approach, the spacecraft moved with the finesse of a falcon, employing its precision thrusters that consumed only three ounces of fuel in a mere eight seconds. Its velocity was now measured in inches per second.

OSIRIS-REx flew over Bennu, its cameras stabilized by the reaction wheels, sweeping back and forth to scan the jagged, boulder-strewn surface beneath. Communication with mission control was crucial, so the spacecraft kept its antenna pointed toward Earth. As it felt the pull of Bennu's gravity, its transmitted radio frequency shifted minutely, like a bird call fading into the distance.

For sixteen days, the spacecraft danced across the surface of Bennu, gliding with quick, precise maneuvers. After three passes of the north pole, it flew along the equator and then dipped down to the south pole. As a new year dawned on Earth, OSIRIS-REx ventured close enough to Bennu to be captured into orbit. To counteract the constant push of the solar wind, OSIRIS-REx positioned itself at the boundary between day and night, known as the terminator. Here, the balance between the force of sunlight and Bennu's gravity allowed the spacecraft to maintain its orbit within the same plane over time. It circled at a leisurely pace, taking sixty hours to loop around the asteroid. In this stable configuration, the spacecraft's future path was predictable, like a well-choreographed dance.

Back home, those of us in mission control erupted in cheers, celebrating this world record–setting feat. Inching around the asteroid at a snail's pace, OSIRIS-REx's first orbit marked a leap for humankind. Never before had a spacecraft from Earth circled such a small space object. Now, OSIRIS-REx was orbiting Bennu about a mile from its center, closer than any other spacecraft had orbited its celestial object of study. This short distance was necessary to keep the spacecraft locked to Bennu, which turned out to have a gravity force only five-millionths as strong as Earth's, the same acceleration experienced by astronauts on board the International Space Station.

Since we were in a whole new realm of operations, Orbit-A belonged to the navigators. For the next two months, they would focus on the spacecraft, free from distractions by nagging scientists, eager to scan their favorite boulder. Setting a new world record comes at a cost. We were in unprecedented terrain and unseen forces from the Sun, Bennu, and the depths of the solar system were pushing OSIRIS-REx around. The navigators needed time to accurately calibrate all of these celestial influences and incorporate them into their orbital models. The intent was to stay in orbit around Bennu through mid-February. The next science phase would begin after that, with a well-rested and ready team.

Bennu had other plans.

One week later, the science team gathered in Tucson as I stood on the riser at the front of the auditorium summarizing our latest results. Suddenly, I heard a brief gasp of surprise off to my right. Glancing over, I pinpointed the source of the noise to be Carl and shot him a questioning look. He responded by widening his eyes and pointing at his screen. I quickly called for a break and beelined over to him. My mouth fell open as I glanced at the scene that appeared on his screen.

Bennu's surface had just exploded.

Despite the presence of over one hundred Osirians, the auditorium was silent.

It was a NavCam image. Bennu was completely overexposed and appeared as a massive white blob in the lower-right corner. It was taken with a long-exposure time. The navigators acquired such shots regularly for their triangulation. By collecting light for five seconds, the images revealed dim stars in the cosmic background. Just like ocean-going navigators in centuries past, they relied on stars to calculate the spacecraft position in the solar system relative to the Earth, the Sun, and Bennu.

"I was flipping through some recent shots for the comms team—and I saw this." Carl stabbed his finger at the screen, circling a cluster of dots just off the limb of the asteroid. "My first thought was, 'Huh, I don't remember *that* star cluster.' I only noticed it because there were two hundred dots of light where there should be about ten stars."

"You're sure this isn't some astronomical trick, like last month?" I replied, referring to a few heart-stopping hours when some faulty imaging made us believe Bennu might have a satellite.

Carl's face hardened with resolve. He pulled up a second NavCam image, taken seven minutes later. He then set the two blinking from one to the other. Again, he pointed at the screen. Sure enough, the dots were still there and were clearly radiating away from the limb of the asteroid. Bennu was ejecting dust particles into space.

First, we had discovered an asteroid covered in giant boulders—and now its surface was exploding? This was not the plan. After a stunned moment, I directed a question to Carl: "Do we have any idea of the cause of the event?"

"We are going to have a busy night," Carl replied. "The cause is unknown at the moment, but we'll get to work right away to formulate some hypotheses. It may be related to the fact that Bennu is near perihelion," he said, referring to the place in an orbit when the asteroid is closest to the Sun.

The next morning, the team gathered once again. The agenda was projected on the screen at the front of the room with a title that blared:

CSI Bennu: Is it safe to remain in orbit?

Overnight, three independent groups had performed impact-probability assessments. Although one of the groups had dubbed their analysis the "hemispherical shell of hell," all three teams came back with similar estimates; there was less than a 1 percent chance of one impact per month in orbit. Given the turtle-like pace of both the spacecraft and any potential satellite, the risk to the spacecraft appeared to be minimal.

I concluded that these particles did not compromise the spacecraft's safety. Microgravity had played with our minds again. What looked like a dynamic explosion was, in fact, the equivalent of crumbs flaking off a broken cracker. Plus, it seemed to be a one-off event. We would stay in the current orbit while we further assessed the risk of impact with one of these objects. There simply wasn't enough evidence of threat to call a time-out, one that would set the mission schedule back by weeks at best.

For the longer term, we needed to decide on how to augment observations, beginning in two weeks, with observations from the other instruments. Gone was any dream of a nice leisurely Orbit-A for the nav team. Where once they had hoped to practice flying, free of any science-team distractions, they now found themselves in an intense operational mode, with uncertainty around every corner.

As soon as the new observations started, we saw Bennu erupt again. This particle ejection event was similar in appearance to the first, and again, the particles were small— this time the size of golf balls—and moving slowly. The threat to OSIRIS-REx might have remained unchanged, but my stress level was at an all-time high.

Almost like clockwork, a third eruption happened two weeks later, providing us with additional data to try to figure out what in the heck was going on. All three explosions originated from different locations, the first in the southern hemisphere, and the second and third near the equator. All three took place in the late afternoon on Bennu. And all three were too minor to damage OSIRIS-REx.

Bennu had a dynamic environment; it popped like popcorn! A big event would send out a swarm of particles. Some would zip away into interplanetary space. Others were lazing around, flying over Bennu for days before reimpacting the surface. It was mesmerizing—and unlike anything any of us had dreamed of. We were the first people in history to observe an active asteroid up close and personal. What had started as a crisis eventually morphed into scientific curiosity.

We developed three working theories, possible mechanisms for the ejection events: meteoroid impact, steam release, and thermal stress fracturing.

Meteoroid impacts are common in Bennu's deep-space neighborhood, and it turned out that Bennu should be hit at least daily by an object with the energy observed in our escaping particles. It became clear that these small fragments of space rock were likely hitting Bennu, possibly shaking loose particles as flaky fragments of their impact. Because of the asteroid's microgravity environment, it wouldn't take much to launch an object from Bennu's surface. The timing of the events was also consistent with the timing of meteoroid impacts. Like bugs hitting a windshield, we expected most impacts to occur on the leading edge of the asteroid as it zipped around the Sun, exactly as observed.

Water release could also explain the asteroid's activity. When Bennu's waterlocked clays are heated, the water could begin to escape the crystal lattice and create pressure. It was possible that steam building up in cracks and pores could agitate the surface, causing particles and gas to erupt. Such mini cometary outbursts could represent the last gasps of what was once a much more dynamic system.

The third theory was a phenomenon called thermal

stress fracturing. Bennu's surface temperatures vary drastically over its 4.3-hour rotation period. Although it is extremely cold during the night hours, the asteroid's surface warms significantly. Rocks expand when sunlight heats them during the day and contract as they cool down at night. As a result, rocks may begin to crack and break down, and eventually particles could be ejected from the surface. This theory gained immediate traction when someone pulled up a video from a group of researchers in California. In August 2014, during the hottest days of summer, they witnessed a granitic dome in California spontaneously exfoliate, with extensive cracking and rock ejection, including an 18,000-pound sheet that popped up straight into the air.

But nature does not always allow for simple explanations. More than one of these possible mechanisms could've been at play. Thermal fracturing could be driving the dehydration, which makes it easy for meteoroids to chop the surface material into small flakes and launch them into space.

Without more data, we were unable to further narrow down these hypotheses. What was clear, though, was that Bennu was a dynamic world, full of scientific surprises. These particle-ejection events showed us that the surface was constantly overturning, revealing the fresh material underneath. With our main objective being the return of pristine material from the dawn of the solar system, this constant resurfacing meant that the primordial organics we were after were likely to be unaffected by long-duration space exposure. It was looking more and more likely that the secrets to the origin of life might be waiting for us on the surface.

Over the next eight months, we observed and tracked

more than three hundred particle-ejection events. We watched over six hundred particles flying around the asteroid, oodles of natural satellites, when earlier we had been consumed with finding even one. The vast majority measured less than an inch and moved at a pace similar to the spacecraft, about as quickly (or slowly) as a beetle scurrying across the ground. On average, we saw one to two particles kicked up per day, with much of the material falling back onto the asteroid. Of the observed particles, some had suborbital trajectories, keeping them aloft for a few hours before they settled back down, while others flew off the asteroid to go into their own orbits around the Sun. In one instance, we tracked one particle as it circled the asteroid for almost a week. The spacecraft's cameras even witnessed a ricochet. One particle came down, hit a boulder, and went back into orbit.

Bennu was literally throwing curveballs—even if they were moving too slowly to hurt the spacecraft, it didn't bode well for the journey ahead.

What other surprises might the trickster asteroid have in store for us? I wondered.

While I waited to find out, it was time to turn our attention back to the giant boulders standing in our way.

Once the ejected particles had gone from crisis to cool science experiment, my thoughts focused on one thing—sample site selection. Data from our Preliminary Survey provided only tantalizing clues as to the locations of potentially sampleable terrain. The higher-than-expected density of boulders meant our plans for sample collection needed to be adjusted. The original mission concept was based on a sample site that was hazard free. However, because of the unexpectedly rugged terrain, it was abundantly clear that such a safe space did not exist on Bennu. I stayed up late each night, hunched over my computer, scrutinizing every image that came down from the spacecraft. I had begun to identify patches of potentially smooth terrain that were, at best, only thirty feet across. Most of the sites were half that size. To make matters worse, such small features were hardly visible in our current data. I felt like I was studying the map without my glasses, and I was struggling to squint my eyes and bring it all into focus.

Once again, I returned to my three pillars of decision-making: deliverability, safety, and sampleability. Based on our current capability, we needed to find open patches of sandy regolith at least 160 feet wide. The initial glimpses of the surface had convinced me that there were no such areas on Bennu. By this stage, my assessment was that the widest hazard-free zones were about ten times smaller than what we had planned for.

We were going to have to increase our deliverability to target the smaller sample sites that had to be there in between the boulders. We needed to switch from laser TAG to *bull's-eye TAG*. Arlin agreed that it was time to summon the software brigade. We were going to need Natural Feature Tracking after all.

CARBON EXPOSED

THE WANDERING CARBON TWIN LAY trapped in its carbonate vein, content to remain there forever. Still, the asteroid belt was full of frenzied motion, stirred by the gravitational fields of the giant planets. Most of the time, they drifted through space, missing each other by millions of miles. But every once in a while, two objects ended up on a direct collision course.

Three and a half billion years after it formed, the asteroid bearing the wandering twin was shattered when Jupiter launched another body its way. The resulting catastrophic disruption broke these ancient worlds into countless fragments, ranging in size from tiny dust grains to massive boulders. Amidst the chaos, the carbonate vein remained intact.

Buried for eons, it was suddenly reexposed to the harsh reality of deep space, peeking out from between two protective blankets of rock. The carbonate-bearing boulder tumbled and swirled in a cloud of asteroid debris. Like a beacon, the bright white mineral reflected light back into space whenever it caught a glimpse of the Sun.

The carbon atom once again felt the familiar pull of gravity as the fragments slowed, stopped, and reversed course. They collapsed around it, forming a spinning pile of rubble, with just enough self-gravity to hold itself together. The wandering carbon twin was adrift once again, encased in a new world, asteroid Bennu. Sunlight heated the asteroid's surface, imparting a small thrust and sending it into the inner solar system. As the carbon atom approached the brilliant blue planet called Earth, it could sense its long-lost twin.

CHAPTER 11

JUNE MADNESS

In early 2019, the OSIRIS-REx spacecraft departed from its orbit to conduct a comprehensive survey of Bennu. The previous flybys had been cursory inspections, providing just enough information for operational planning. However, this time, the mission was focused on science, and we aimed to produce the highest-resolution maps of any celestial body ever made, including of the Earth.

OSIRIS-REx moved with unprecedented precision and control around Bennu, executing sharp turns and defying gravity with superheated bursts of fuel. As agile and effortless as a hummingbird, the spacecraft flew up, down, forward, and backward, making sudden starts and stops with ease—a graceful, synchronized flight through space.

It conducted fourteen of these sweeping passes, one per week, allowing it to study geological features at various latitudes and times of day. The giant boulders on the surface cast elongated shadows and appeared fractured and fragmented. The craters were small dimples, a testament to the constant cosmic bombardment that slowly breaks down the spinning rubble. Each bit of information was diligently transmitted back to Earth for analysis. The data it collected were both scientifically groundbreaking and breathtakingly beautiful.

Night after night, I pored over the information coming down from deep space. With each dispatch, dread spread through my core. Despite the improved data quality, most of them still contained nothing but giant boulders, revealing a terrain so challenging that I balked at the notion of sending our magnificent spacecraft down into the mess. Would we really be able to find a spot to sample?

Even if, or when (as I forced myself to use positive and upbeat language) the engineers solved the bull's-eye TAG challenge, when the TAGSAM head contacted the surface of Bennu, it must do so in a safe area. The last thing we wanted was a successful contact, only for the spacecraft to be damaged beyond our ability to bring it home. Additionally, the site had to contain abundant pebbles and sand-sized particles that would allow TAGSAM to scoop up two ounces or more of regolith.

Prior to launch, the engineers had developed a purely mathematical way to find such sweet spots on Bennu. Their algorithm spit out a product they called the "Treasure Map." This Treasure Map algorithm ingested variables like the tilt of the surface and the average grain size. Of course, figuring out the grain size was a conundrum, since we now knew that the thermal data didn't provide an accurate measurement. At that point, the only Bennu particles we had found that TAGSAM could ingest were the ones we watched explode from the surface. That these cosmic bullets were of the right size gave me some comfort. At least I knew they were there somewhere.

As I agonized over Bennu's ultimate challenge, our colleagues on the other side of the planet were moving forward with their own asteroid mission. Japan's Hayabusa2 spacecraft mounted its first sample attempt at asteroid Ryugu. When the spacecraft's sampler touched the asteroid, a small bullet was fired into the surface. The resulting ejected materials were collected by a catcher at the top of a funnel-like device that directed shattered fragments into a return capsule. The whole interaction lasted milliseconds, making our five-second sample collection seem like a lifetime.

We gathered in the Drake Building to watch our friends in Japan execute this delicate maneuver on the other side of the solar system; their *moment of terror*. Even though I had witnessed dozens of these types of operations, each was as nerve-racking as the first. Images of Mars Observer, Polar Lander, Genesis, Philae, and Hayabusa sprang to mind. I steeled myself to watch yet another group of friends suffer defeat at the hands of our cruel solar system.

This time, though, everything went perfectly. Preliminary data from the spacecraft indicated that a sample of Ryugu had been safely stowed on board. Chills washed over my body along with pure joy—finally, someone had cracked the asteroid-sampling challenge. The concept had been proven. My friends had triumphed. A wide grin spread across my face as I looked around the room. Everyone was cheering and high-fiving each other. Score one for the humans!

Luckily for us, Hayabusa2 got some amazing views of Ryugu during their sampling. There were clearly particles liberated from the surface as a result of the spacecraft's contact. It was not apparent if this was from the physical act of the touchdown, the thruster firings, the bullet, or some combination. One thing I knew for sure: this boded well for OSIRIS-REx at Bennu.

As we studied the data from Hayabusa2, it was evident that the sampling event had changed the asteroid's surface. Ever vigilant, Arlin was concerned that the surface had been modified from Hayabusa2's back-away thrusters. He postulated that either the back-away thrusters had blown off the "sunburned" surface layer, exposing darker regolith beneath, or the hydrazine fuel interacted with the top layer somehow. He was particularly focused on making sure that our MatchPoint rehearsal, the final maneuver before TAG where we would approach within a hundred feet of Bennu, did not disturb the "pristine" surface. The last thing we wanted to do was contaminate our site right before sample collection.

I contacted my colleagues in Japan and asked them to send over some of the data from their rehearsals for us to review. As we clicked through the images, it became obvious that the thrusters not only modified the surface, but also caused some boulders to roll away from the impingement sites. Arlin committed to a thorough study of possible effects from our spacecraft, leading to a potential redesign of our rehearsal strategy.

Five days after Hayabusa2's success, we held the first official meeting of the Site Selection Board. There were representatives from science, TAGSAM, spacecraft, and navigation. Arlin was there from Goddard, along with watchers from NASA Headquarters. These twenty people were charged with agreeing on the best site on Bennu to collect our sample. Imagine the challenge that two people have just deciding where to go to dinner, and you get a sense of the chaos that was about to ensue.

Though I had spotted a handful of locations using the Preliminary Survey data based on simple gut instinct, the engineers were convinced that their Treasure Map algorithm was going to lead us to our science gold mine. The Treasure Map was based on the concept of searching for flat areas using the asteroid shape model, wherein one uses a series of triangular facets to model the shape of small body. The higher the resolution, the more facets needed to tile across the shape. With the inputs we had at this point in our exploration, the Treasure Map algorithm identified twenty sites across the surface. Surely one of these would be the perfect location to collect our sample.

My euphoria was short-lived. On the shape model, these sites looked great. They were covered by the small triangles, each one nearly parallel to the other. To truly determine if they were good sample sites, we needed to look at the actual images. The computer model would only take us so far. One by one, we pulled up each Treasure Map location. It felt like déjà vu. I had been studying this landscape for months, staying up late each night as the spacecraft beamed its observations back to Earth from across the solar system. All of these sites looked horrible. Each one was covered in big, blocky boulders, many of them well over three feet across, fifty times the size of the biggest particles TAGSAM could ingest. The shape model was simply too coarse for the Treasure Map algorithm.

As we clicked through the last images, it became apparent that we would be leaving this meeting without a single potential sample site identified. I scanned the faces around the room and watched as each member of the board came to this same conclusion. The clock was ticking, and the navigators were anxious for guidance. We needed to provide them with some example sites so they could begin detailed planning of the complex series of maneuvers required to get us accurately and safely to the surface. At this point, I had nothing to tell them.

The answer was a call to arms. I put out the word to the entire team. Not just the science team, the *entire* team. I told them to start scrutinizing every bit of data. My best lead was still those sites I had found by simply poring over hundreds of views of Bennu's surface. I knew we could harness our collective power to do the same. The Treasure Map had failed; now we would try good old human intuition. After all, there is no computer on Earth better than the human brain.

Everyone sank their teeth into this task with vigor. Many of them followed my lead and started looking at image after image as they came down from each subsequent Detailed Survey campaign. They identified a dozen additional locations. Then, we turned to the public. We began releasing data on the CosmoQuest website, a community of citizen scientists interested in planetary exploration. Enthusiastic users performed a systematic visual inspection of the entire asteroid, which yielded over a dozen more possible locations. We also trained a machine-learning algorithm to find areas missed due to human error and found several more possible locales. Six weeks later, in mid-April, we had fifty possible sample sites on Bennu.

When I sent the list of fifty to the science team, their morale plummeted. They were standing ready to compile detailed dossiers on up to twelve potential sites. The sheer number of options overwhelmed them.

"What, exactly, are we supposed to do?" one member queried in the chat room.

"Correct me if this is wrong, but what this plan amounts to is us ranking the sites based on our gut feeling of the probability that we'd obtain two ounces of sample if we successfully TAG in that spot," chimed in another. "And by 'that spot,' we mean, what? A circle chosen to exclude as many visible rocks as possible?"

Mike Nolan, who was serving as our science team chief, tried to assuage concerns: "I see this task as going through these and trying to identify which sites may have fooled us by, say, being smooth except for that big rock in the middle. The goal now is to narrow down the sites that need significant effort on further measurements or modelling. We're also getting to be pretty familiar with the sites, which I think is important. A lot of eyes have now seen the surface, and I am reasonably confident that all plausible sites are on the list. Please, go through the suggested sites and 'stoplight' triage them: green, yellow, red, based on your visual inspection. At the board meeting next month, we will rank them."

My unease spiked when our NASA Headquarters representative spoke up: "This whole system seems to be based on people's instincts. The process needs to be specific, as quantitative as possible…and documented. Remember, you are going to have to stand up in front of the associate administrator at Headquarters and defend your decision. Based on what I'm seeing here, he will toss you out of the room."

With that chaotic guidance, the Site Selection Board gathered in early May to triage our options. Over ten long hours, we went through the massive list of fifty sites, each board member voting red, yellow, or green as they came up on the screen. Only twelve were red. Even with that consensus, the board still refused to eliminate them entirely from consideration. They had reached a state of analysis paralysis.

As I walked back to my office, my angst started to build. All the board had done was a first pass of classifying sites — as individuals, without guidelines. What we needed was a well-conceived, documented method, instead of this seat-of-the-pants popularity vote. This decision being deemed mission critical, NASA Headquarters was keenly interested in our progress. In fact, I was scheduled to brief senior NASA leadership two weeks from now. I could not, with any confidence, defend this method in front of them.

I spent my nights obsessing, tossing and turning while butterflies roiled my stomach. Clearly, I needed to get this situation under control. What we needed was someone who knew the reality of spaceflight and the magnitude of the decision that we were making. I turned to the one person I knew who could bring some discipline to this motley crew — Heather.

The rocket launch was a natural transition point for many team members. Some of them simply preferred to build spacecraft; others achieved recognition for their efforts and were offered lucrative positions elsewhere. And some just wanted to go out on a high note. Ed fell into that last category. He was near the end of his career, with many accomplishments to be proud of. Now he was looking forward to a more relaxed and leisurely lifestyle.

With Ed moving on, I decided to promote Heather to the deputy PI position. At first, this nomination was met with howls of protest from NASA.

"She is not a scientist!" they wailed.

"How can she serve as the deputy without a PhD?" they moaned.

"For goodness' sake, she only has an MBA," they condescended.

But the last thing I needed was another scientist. The mission was awash in them, and they were trapped in indecision. I had the science under control.

No longer content to wait for this board to deliver a smaller set of sites for my consideration, I asked another of my superstars, Dani, to have her team put together a catalog of all the current sample sites. Dani's eyes lit up at the assignment and she broke into a bright and open smile. True to form, she tackled this challenge with the grace and persistence that was the hallmark of her pursuit of scientific knowledge.

Her team got to work measuring the size of as many of the rocks as possible. While we did not have enough resolution to see the one-inch particles that TAGSAM could collect, we could certainly see particles ten times that size. Even more importantly, we could clearly see where such particles were not. These areas looked "unresolved," meaning they were generally featureless with no obvious boundaries. If there were fine particles on the surface, that is exactly what they would look like.

As I flipped through the resulting catalog, it became clear that we were spinning our wheels on some of these sites. Many of them were covered in boulders way too large for TAGSAM to collect. I corralled two key team members — Anjani Polit and Carina Bennett—to help whittle them down. Anjani was our system engineer, the person responsible for ensuring the mission met its requirements. Carina was a top software engineer, the one who compiled all the images in the catalog and led the rock-counting efforts. At this point, nobody knew the data better than she did.

"This ends now," I mandated. "The three of us are going to sift through all fifty of these sites, just the three of us, and narrow them down to a manageable number."

We quickly eliminated all sites that had at least half their area covered with yard-size boulders. We also removed some sites that were near others that clearly looked better. We narrowed them down to a group of top priority sites, which I dubbed the "Sweet Sixteen."

The next day, I sat down with Heather to coordinate a strategy. She was confident that she could organize the team around these locales and provide a quantitative, well-documented process to get us to the primary sample site. Queuing off my Sweet Sixteen moniker, we intended to systematically cull the sites down to eight, then four, then two, leading to our champion.

And just like that, we were in the middle of our very own June Madness. For the first time in months, I slept a full eight hours. I woke up confident, ready for the tournament to begin.

The following week, the newly inspired Site Selection Board convened.

"Listen up everyone," Heather began. "I am convinced that our sample site is in the Sweet Sixteen. Our task right now is both to lay out a path to selecting the Final Four and also to relieve the enormous stress that we have put the operations team under with our dithering."

I scanned the faces around the room, making eye contact with every individual. It may have been my imagination, but as I locked eyes with our navigation team chief, I saw a bit of relief in his eyes.

"Navigation needs a definite target for the first reconnaissance pass," Heather continued. "These passes will be incredibly challenging feats of aerodynamic acrobatics. To collect the detailed information we need, the spacecraft needs to perform a series of precision maneuvers to get directly over the regions of interest. Only then can the cameras gain enough resolution for us to see if there is sampleable material present. So, the navigators need to know exactly where we want them to go, and they needed it yesterday. By the end of this week, I want half the remaining sites eliminated. It is time to move on to the Elite Eight!"

This mandate seemed to intimidate the team. They were so hesitant to commit to a particular location; we needed to lift their spirits. "To get started," I added, "I have a surprise. Has anybody here ever heard of the rock band Queen?"

The look of puzzlement that swept the room was worth the wait. I had been holding this revelation in reserve for a while.

To add to their bewilderment, I started passing around sets of bright red stereo viewers. These devices, called Owls, were modern-day versions of the View-Master toys that were wildly popular when I was a kid. By dropping in a circular wheel that contained a series of stereo images, the world suddenly popped into a three-dimensional wonderland.

"It turns out that Brian May, legendary rock guitarist, is also a renowned asteroid scientist. He has developed a system for viewing asteroid data in three dimensions using his Owl Stereoscope 3D viewer, which works with your cell phones. And…he's produced pairs for every one of the Sweet Sixteen sites."

Seeing those sites pop out in three dimensions made a real impact. They also made the challenge that we had been wrangling with eminently clear. Our requirements were in direct conflict with each other. The safety criteria demanded nice, flat patches of asteroid. Sampleability needed abundant small particles for ingestion into TAGSAM. As I scanned through the stereo images, it became obvious that the small particles were concentrated in impact craters, tiny scars that recorded cosmic collisions over Bennu's history. Unfortunately, they were all bowl shaped with steep walls, which represented a substantial risk to the spacecraft. Deliverability was not happy—the spacecraft team was still furiously working out how to improve our accuracy to hit the tiny bull's-eyes that we were nonchalantly drawing across our potential sites. Inspired by our musical colleague, we all started singing the "Bennu Site Selection Blues."

A few days later, the board reconvened. Given the enormous uncertainty with all of the potential sites, Heather directed them to categorize each location. We both agreed that we wanted the Final Four to represent diverse terrain types. This approach was a means of hedging our bets, placing markers in four very different areas. Working as a group, the board divided the Sweet Sixteen into three different categories: the southern terrain, salt and pepper, and dark red craters.

The southern terrain sites were located in the southern hemisphere, on the wall of Bralgah Crater, the giant impact scar that dominates the lower half of Bennu. Its walls have a texture that is distinct from the rest of the asteroid, with clear patches of smooth terrain interspersed with larger rocks. The sites that we identified along this region included some of the flattest areas on Bennu. Even if we had to target a site with some boulders in the way, we felt confident that the spacecraft could make it to the surface and retreat undamaged. We might only collect some dust grains, but at least we would make it out unscathed. Three sites were in contention here, all of which were right next to each other, like a string of pearls along the neck of the giant basin.

The salt-and-pepper sites were scientifically compelling. Bennu's surface is striking in its combination of both bright and dark rocks. These sites appeared to have a nice mix of both rock types, providing the best opportunity to sample the diverse components that make up the asteroid surface, allowing us to unravel its geologic history. Four sites were in this category. One of them quickly rose to the top of the list, as it looked reasonably safe, meaning that there were no building-size boulders in its immediate vicinity. It was also one of the locales from my initial survey back in December.

The main concern with this site was the area at the center of the crater. This region appeared to be a mound, characteristic of much larger craters on the Moon and other planets. This hump resulted in surface-tilt values that the engineers had deemed unsafe. It was also much darker than the surrounding material and the scientists were concerned that this region was heavily altered by impact, making it less scientifically valuable. As I listened to the board debate, I saw possibility. I had learned from my years in Denver that engineers always stuff their pockets with margin — capabilities that vastly exceed mission requirements — as a way to guarantee mission success. I just needed to pick their pockets and release that margin.

Next up were the dark red craters. Bennu, being a B-type asteroid, appears blue when looked at through a telescope. Most of our cameras produced gray images, since they let in all wavelengths of light to create the strongest signal. The MapCam, in contrast, had a set of color filters that allowed us to view Bennu in its full Technicolor glory. We could drop in the four filters, one at a time, and observe Bennu in a distinct color: blue, green, red, and infrared. When Dani applied this technique to the dark red craters, they popped out from the background's rocky nightmare. By combining the color data, Dani was able to reveal the diversity of Bennu's surface. The darkest, finest-grained material on Bennu, such as that found in these small craters, appeared red. Even more importantly, the size range of this crater population matched our predictions for impacts over the past 100,000 years, a blink of an eye in terms of Bennu's billion-year history. These small craters thus represented the freshest sites on the surface, areas that had only recently been excavated by cosmic collisions. As she presented her conclusions, I had to wipe the drool off my chin. We had to get into one of these tight spaces.

To compare the sites, Heather had the board establish four quantitative metrics. The first was the total area of the site, since wider areas were better for deliverability. Next, they looked at the fraction of unresolved material, which provided a proxy for sampleability until we could get higher-resolution data. For safety, they quantified the tilt at the center of the site, which was still driven by the ultraconservative (in my opinion) approach. Finally, they looked at the calculations from the navigators on the likelihood the spacecraft would reach the targeted destination.

For the first choice, we focused on the southern terrain. Given their location on the wall of a crater, they were relatively flat and therefore their deliverability and safety scores were the highest. One of the three sites appeared to have a much lower boulder density than the other two. Staring at these scenes, I imagined the geologic events that had resulted in this relatively hazard-free area, there on the southern rim of Bralgah Crater. A boulder was perched near the top of the rim. It looked like material was tumbling down the wall and piling up on the crater floor. The massive stone at the top of the rim seemed to be blocking this material, creating a runway-like strip down the wall that was relatively free of hazards, like a celestial bodyguard preventing the biggest rocks from entering the site.

The board selected that site as the first location for detailed reconnaissance. With its promotion to the tournament semifinals, we decided we could no longer keep referring to sites by a coldhearted numerical designation at that stage. Just like naming Bennu, providing locations with a true moniker would make them more real. We named all the Final Four sites after birds native to Egypt, since

Bennu was often represented as an Egyptian heron. Due to its southern location, we wanted it to begin with an *s*, so Sandpiper it was. At the end of the day, we were down to the Elite Eight, with Sandpiper getting a pass straight into the Final Four.

CHAPTER 12

THE FINAL FOUR

ON JUNE 12, 2019, OSIRIS-REx took another leap into the unknown and broke its own world record for the closest orbit of a celestial body. The spacecraft swooped down to only four-tenths of a mile above the surface of Bennu, marking the start of a new chapter in the mission, known as Orbit-B.

OSIRIS-REx flew along the terminator, navigating the delicate dance between Bennu's minuscule gravity and the relentless push of the solar wind. The spacecraft maintained a constant lifeline with Earth during this phase, with the Deep Space Network antennas tracking its motion and measuring both its range and velocity like a watchful guardian. The data collected showed a slight wobble in the spacecraft's path, allowing us to trace out Bennu's weak gravity field. The field reflected the internal structure of the asteroid, revealing that its core was less dense than its outer layers, with void spaces located at the center. Like a child on a merry-go-round, the centrifugal forces were pushing material away from the asteroid's axis.

As Bennu rotated beneath the spacecraft, a small mirror swept a laser beam across the surface. With just a single pulse, the onboard laser altimeter was able to take a three-dimensional snapshot of a patch of Bennu, a dynamic virtual terrain for us to explore back on Earth. Early in the mission, the laser collected one hundred data points per second. With Bennu now so close, its sampling rate increased to ten thousand pulses per second, covering the surface with topographic maps detailed down to a fraction of an inch. As the scans progressed, the full extent of the surface roughness came into focus, with towering boulders resembling abandoned skyscrapers scattered across a desolate and ancient landscape.

What seemed like ages ago, but really was just two short months in the past, the Treasure Map had failed owing to the coarse resolution of the shape model. The new laser data allowed us to improve that resolution by a factor of over two hundred. Now we could calculate dozens of tilt values underneath the TAGSAM head alone.

I spent countless hours wandering around Bennu in virtual space, reveling in viewing every nook and cranny in glorious 3D. I roamed across the rough and boulder-strewn terrain, admiring the jagged rocks scattered across its surface. I dove into the craters, caused by impacts from other asteroids and space debris. The diversity of the surface was a geologic marvel, and I caught glimpses of areas of Bennu's surface that were coated in a bright material that reflected sunlight, while other areas were shadowed and appeared much darker. Leaving the virtual world took a massive effort of will; but it was time to get back to business.

Why the hell did we wait so long to get this data? I wondered to myself.

The reason, of course, was that we had built our mission around a cautious, step-by-step approach to getting ever closer to Bennu. There were too many unknowns to go rushing into a deep orbit right away. The particle ejection events had shown us that these spinning piles of rubble didn't behave like we expected them to. No doubt, Bennu still had some surprises in store for us.

The team devoured the data like a bunch of starving beasts. This was what they had been waiting for. With this information in hand, we were able to resolve a lot of uncertainty surrounding the safety of the sites. And by safety, we meant tilt. Our biggest worry was that the spacecraft would hit the surface at an angle, causing it to stumble, tumble, or lean over and collide with a giant boulder. There were two ways out of this conundrum. Either improve the bull's-eye TAG accuracy, allowing us to zero in on the smoothest location possible, or release some spacecraft margin, allowing contact with steeper tilts than originally built into our plan. Fortunately, the programmers had been busy, and they had driven the performance to an accuracy of less than sixty feet, more than a twofold improvement over the original design. This value was still large compared to our sample sites, but it was progress.

We combined the stunning topographic maps with the improved understanding of spacecraft guidance capabilities to make our decision on the last three candidate sites to enter the Final Four. The navigators constructed maps to demonstrate the likelihood of a successful TAG attempt in each of the remaining sites. These maps were rudimentary, a spray of red and green dots on top of each location. Each dot represented the result of a single "Monte Carlo" simulation. These calculations, as the name implies, are basically games of chance. The actual path that OSIRIS-REx would take on its way down to the surface would be the result of many different factors—things like the direction and timing of our departure from orbit, subtle variations in Bennu's gravity field, and the force of the thruster firings during the final moments before contact.

We could only account for these factors so much. Each variable had an associated probability distribution—essentially the likelihood of each unknown quantity achieving a specific value, like rolling two dice. The most likely outcome is rolling a seven. But we needed to know what would happen if Bennu gave us a series of snake eyes and boxcars—low-probability events with potentially catastrophic consequences. For each run of a Monte Carlo simulation, these distributions were sampled randomly, and each dot on the map represented one unique combination of values for each variable in our equations. The dots were color coded: red dots indicated landing on a spot with a tilt value above the spacecraft safety limit. Green cases would successfully tag the surface.

For each site, four numbers were presented: deliverability (the fraction of the Monte Carlo dots that were green), tilt at the center of the site, percentage of the site covered by unresolved material (which we assumed meant small, sampleable particles), and the science value (our confidence that the material contained water-rich clay minerals and carbon-bearing organic molecules).

The deliverability maps were the most striking. The map of the first site had a tiny cluster of green dots at the center of a sea of red. The second area wasn't much better, with the number of green dots growing to about a third of the total. My heart sank as these images flashed across the screen. *Did we really have only a 30 percent chance of making it to the surface?*

The next two maps changed my perspective. The first showed the result using the current spacecraft specification for tilt safety. The site was covered in about half red and half green dots. Then, we saw the results assuming we could relax the tilt requirement by a meager five degrees. Almost every point was green. I inhaled a deep breath. For the first time since arriving at Bennu, we had a clear demonstration of one case where we could predict, with an extremely high degree of confidence, safe contact with Bennu—if we could only relax the tilt requirement those few degrees.

This information made the final selection straightforward. NASA Headquarters representatives were in Tucson for our deliberations, and I watched with relief as they nodded their heads at each discussion point. By the end of the day, we had made our decision. In addition to the previously chosen Sandpiper, we selected one site in a dark red crater in Bennu's northern hemisphere, which we nicknamed Nightingale (*n* for north). The other two sites were of the salt-and-pepper variety and were located near Bennu's equator. One became Kingfisher and the other Osprey. As I reviewed the Final Four, I couldn't help but smile. Nightingale and

Osprey were two sites I had picked out eight months ago, using those fuzzy images from our early days at Bennu.

Sandpiper, Osprey, Nightingale, and Kingfisher were locked in—and so was our schedule. For the next five months, we would have our feet on Earth while our minds were on Bennu, analyzing, discussing, and dreaming about small patches of the asteroid that, altogether, added up to about the size of a basketball court.

It was time to move on to the semifinals and then the tournament championship, in quick succession. To stay on schedule, we needed to select a prime site by early December. Safety and sampleability assessments would be incrementally improved as we progressed through acrobatic observations during our four reconnaissance passes. This uncertainty meant that we still had substantial schedule risk; we did not know what the Recon Phase would reveal.

This phase was scheduled to begin in October. That gave us a couple of months to catch our breath, regroup, and get some science done. After all, the whole point of these deliberations was to find a spot for TAGSAM to harvest some scientific treasure. Since the first spectral data came in back in December, we had known that Bennu was wet, built predominantly out of clay minerals that locked up water in their crystal structure. The real prize, however, was the carbon—pristine organic material from the early solar system. Once again, spectroscopy delivered the goods.

After months of compiling spectral data, we obtained convincing evidence that carbon-bearing material was widespread on the asteroid's surface. Similar to water, carbon chemistry is most apparent in a specific wavelength region of infrared light. The challenge is that it occurs both in organic molecules, which contain carbon in a form often found in biology, and carbonate minerals, like the salty white crusts that build up around our sinks. As the analysis progressed, we concluded that Bennu had both. The organics appeared to be everywhere, with the strongest signals coming from the darkest boulders. The carbonate minerals, on the other hand, seemed to be concentrated in the bright regions.

These carbonates provided the final clue to understanding Bennu's salt-and-pepper texture. As Dani and the imaging team worked systematically across Bennu's surface, they distinguished two main types of rocks: dark and rough, and the less common bright-and-smooth variety. Not only did the boulder types differ visually, but they also had their own unique physical characteristics. The thermal analysis team, eager to prove their worth after the rocky horror revealed during approach, reported that the dark boulders were weak and highly porous. Thus, their thermal properties were very much like that of beach sand. This eye-opening result was why we were fooled by the telescope data. These dark boulders were also friable, meaning they likely would not survive the journey through Earth's atmosphere. A thrill shot through my body—it looked increasingly likely that OSIRIS-REx would return extraterrestrial material never before seen in any laboratory on Earth.

The bright boulders, on the other hand, seemed much stronger and less porous. These boulders hosted the carbonate minerals. The precipitation of these salts in cracks and pore spaces probably acted like cement, strengthening the boulders, and reducing their porosity. Once all this data came together, some of the features I had spotted on Bennu early in the mission started to make sense. Though I had spent the past eight months ogling the asteroid, I had yet to grow tired of it. As I scanned the boulder-strewn terrain, it often reminded me of hiking in the nearby hills outside of Tucson. I found myself imagining placing TAGSAM on the trail and collecting the gravel and pebbles underfoot. A number of Bennu's boulders had bright white visible veins shot through them, stretching over a yard in length and tens of inches thick. I saw similar veins in the Arizona desert, ones that were also made of carbonates. I wanted to reach out and touch them, run my finger along their silky stripes, just like I did when Carleton turned me loose in the vault at ASU twenty years earlier. All of a sudden, it clicked into place.

Bennu's rocks were made of clay minerals, carbonates, and organic molecules. On Earth, these minerals form in extensive hydrothermal systems, where scalding-hot water boils through miles-long fractures, scavenging elements, hydrating rocks, and synthesizing organic matter. Though we had clear evidence of such reactions in meteorites, the consensus in the scientific community was that the fluid flow was limited to microscopic scales. Such minute elemental transport could not explain the features we were seeing on Bennu. Instead, Bennu's parent asteroid had to have been a giant convecting ball of mud, circulating fluid for millions of years and completely altering the original mineralogy. Although the parent body was destroyed long ago, we were seeing evidence of what that watery asteroid once looked like. What made this even more exciting was comparison to the locales on Earth where this process takes place—deep-sea hydrothermal vents. These underwater geysers are prime candidates for the site of the origin of life on our planet almost four billion years ago.

If we could collect a sample of Bennu, we might be able to provide a crucial missing link to figure out how life formed on our planet. *We had to get the sample home.*

Fortunately, we were making progress on that front. The Final Four selection meeting had given me glimmers of hope. Some of the Monte Carlo calculations came up with almost every point plotting green, but it was the *almost* part of the result that I started to obsess over. When I zoomed in, it was clear that a lot of the red points landed squarely on boulders. What would happen if the spacecraft contacted one of these red zones? The answer was obvious; we would suffer the same fate as the first Hayabusa mission.

As I showed Arlin the latest images of boulders scattered around the Final Four sites, his face set in grim determination.

"Give me some time to work on this," he said.

When we next met, Arlin looked exhausted yet resolute. His eyes were yet again ringed in blue, and he slouched in his chair. I could tell he had been up all night. Despite his fatigue and stress, his sense of urgency was undiminished.

"During TAG," Arlin explained, "OSIRIS-REx will literally be on the other side of the solar system. Any information we get from the spacecraft will be long over. We will simply be reliving its past."

I swallowed hard as the implications sank in, although I had long known this to be true.

"With such vast distances and communication lags," he continued, "the spacecraft will be alone, without us to guide it, and we need to give it some tools to stay safe."

"Right," I replied, "the laser data made it very clear: while these sample sites all contain some broad, flat areas, there are always a few spacecraft-killing boulders scattered about."

"OSIRIS-REx needs to get even smarter, for its own protection," Arlin concluded.

A few weeks later, the engineers had come through. They concluded that it was possible to augment NFT with a new feature, which they called the Hazard Map. This map would need to be loaded into the spacecraft's memory well ahead of TAG. After the MatchPoint, the final maneuver before contact, NFT would continue to calculate the spacecraft's position and predict the location on the surface where it was heading. We color-coded the surface with red areas designating hazards and green areas indicating safe zones. If the spacecraft was approaching a hazardous area, it could wave off the attempt and retreat to safety by firing its thrusters, much like a skilled pilot deftly aborting their approach to an aircraft carrier to avoid a collision.

"The bad news," he emphasized, "is that, just like we saw during the Hayabusa2 sampling event, once we back away, the spacecraft will disturb the surface."

"If OSIRIS-REx fires its thrusters just before contact," my eyes wide as I concluded, "then we could blow a hole in the surface."

"Right!" he replied. "That means we only get one chance to sample at any site. If we wave off, then the surface is disrupted, the natural features that we painstakingly characterized will likely be gone, and our fancy Hazard Map will be rendered useless."

"Right now, the probability of wave-off is hovering around 20 percent, even for the best sites," I recalled.

"Indeed. That means we can't just focus on one site. We need a 'hot' backup, a second site that is ready to go in the event that we blow the first one."

"And..." I slowly continued, "since the science team is responsible for building the natural feature catalog, that means we just doubled their workload."

As soon as I finished with Arlin, I called Heather. "We're going to need some more people, more time, and more money to get ready for this TAG event."

◆ ◆ ◆

We were on a roll. The engineers were busy building a bigger brain for OSIRIS-REx. The navigators were systematically knocking down the uncertainties in their trajectory solutions, and the science team was striking literal pay dirt. And then, improbably, our spirits were lifted further. In addition to upgrading NFT, the engineers had also been working tirelessly to relax the tilt requirement.

In hindsight, the answer seemed obvious. In true conservative engineering fashion, it turned out that their calculations to determine whether or not the spacecraft would trip and fall upon contact had assumed a perfectly slippery surface. We called this the "banana peel" model. With such a slick surface, even the smallest angles could send the spacecraft spinning. Their model was missing friction. Once they applied this basic parameter, they concluded that the tilt requirement could be relaxed considerably. I had been hoping that they could give us an extra five degrees. They released more than twice that amount. With that, the green dots on each graph increased considerably.

We had come a long way over the past four months and had opened up what seemed like vast amounts of accessible Bennu real estate. We had gone from a 50 percent wave-off probability to 1 percent at Sandpiper, our most promising site. I had even started regularly sleeping through the night again.

During this phase, OSIRIS-REx conducted a series of reconnaissance missions, scanning the Final Four sample sites. Departing from its orbit, the spacecraft carefully surveyed the asteroid's surface for hidden treasure troves of information. With each flyover, OSIRIS-REx swooped closer, capturing high-resolution images. The spacecraft acted like a vigilant lookout, scouring the asteroid for signs of danger or potential bounty. The features and landmarks charted during these missions were like buoys and beacons, directing OSIRIS-REx through shallow and rocky asteroid terrain. Using this information, the spacecraft's autonomous navigation system could precisely usher OSIRIS-REx toward its target sample site, much like a lighthouse guiding a ship safely to shore.

Situated in Bennu's southern hemisphere, Sandpiper was the first recon target. It was cherry-picked early because we thought it was the safest and most deliverable site on Bennu. It looked like a long, flat runway descending down the wall of the giant Bralgah Crater. Even though our intuition suggested that this was a steep, slippery slope, we couldn't trust our instincts when dealing with microgravity. All that was required was a shift in perspective, and a shift in the angle at which OSIRIS-REx approached the surface. As long as we could descend along a path that intersected the surface at ninety degrees, it didn't matter what the local gravity field was. It was so small we could simply cancel it out with our thrusters.

As the spacecraft zoomed over the site, the onboard cameras snapped hundreds of images. The data came down over the next few days and I anxiously awaited each bit. As we stared at the most recent shot from Sandpiper, new, previously unresolved features snapped into focus. Even though fine-grained material was present, the sandy regolith appeared trapped between larger rocks. This interspersal of sampleable material and potential hazards would make it difficult for TAGSAM to operate. With this new information, the wave-off probability shot up to almost 30 percent, and Sandpiper no longer looked like the safest site on Bennu.

The following week, OSIRIS-REx flew over Osprey. As we analyzed the data, a strong spectral signature of carbon-

Site Sandpiper seen from Recon A. The field of view is forty-eight feet. Credit: NASA/Goddard/University of Arizona.

rich material emerged, exactly the kind of sample we were hoping to retrieve. In addition, the surrounding region was relatively hazard free. The navigators calculated a less than 6 percent chance of waving off. I liked those odds. However, the high-resolution images of Osprey suggested that the site may be scattered with material too big for TAGSAM to ingest, lowering the probability of collecting a sufficient sample. Here was a site with an almost perfect chance of safely contacting the surface, but the sample seemed just out of reach. Bennu was playing tricks with us again.

Site Osprey seen from Recon A. The crater has a sixty-six-foot diameter. Credit: NASA/Goddard/University of Arizona.

Kingfisher, the favorite of our thousands of committed CosmoQuest citizen scientists, was the target of the third recon flyover. This site was selected because of its location in a small crater—meaning that it may be a relatively young feature compared to Bennu's larger craters, like the one in which Sandpiper was located. Younger craters generally hold fresher material. The new images showed that, while the original crater may be too rocky for OSIRIS-REx to safely TAG, a neighboring, tiny crater appeared to contain the finest-grained material on the entire surface. Its small size, though, resulted in a wave-off probability similar to Sandpiper.

Site Kingfisher seen from Recon A. The small crater has a diameter of only twenty-six feet. Credit: NASA/Goddard/ University of Arizona.

It struck me that we had a Goldilocks problem. Osprey was safe and deliverable, but sampleability was questionable. Kingfisher had a gorgeous patch of very fine particles, but it was so small that we could never guide the spacecraft safely to that location. Would Bennu ever give us a break?

The final flyover targeted Nightingale, which was located in a sixty-five-foot-wide crater in Bennu's northern hemisphere. Nightingale's regolith was dark, and the crater floor was relatively smooth. Because it was located so far north, temperatures in the region were lower than elsewhere on the asteroid and the surface material was therefore expected to be well-preserved. It was also the reddest crater on Bennu, implying that it was relatively young, and the regolith freshly exposed. The site would likely allow for a pristine sample of the asteroid, giving us insight into critical moments of solar system history. Its wide area and abundance of small particles resulted in the best sampleability score. Its spectral signatures contained evidence of both hydrated minerals and organic compounds, giving it the highest science value. My mouth dried up as I brought up the laser data. A giant boulder sat perched on the eastern rim of the crater. This monster had a sharp, peaked profile that was straight out of a Tolkien landscape. I nicknamed it Mount Doom.

With the last of the recon data on the ground, the board gathered in November to make their recommendation to me on this mission-critical decision. For better or worse, it was time to crown our champion.

The board voted unanimously to remove Kingfisher and Sandpiper from consideration. Nightingale and Osprey had made it to the championship game. It was down to science versus safety.

Site Nightingale seen from Recon A. The crater is sixty-five feet in diameter. Credit: NASA/Goddard/University of Arizona.

In my mind, Nightingale had clearly won. This mission was supposed to be all about the sample and its scientific value, and in these categories, Nightingale was way ahead. The spacecraft team at Lockheed and the mission system engineers from Goddard had other thoughts.

Heather called the board to order, and each member spoke in turn. The navigators presented their latest simulation results. With the Hazard Map now in play, there was less than a 1 percent chance that OSIRIS-REx would make hazardous contact with the surface, no matter where we went. It would simply wave off and live to sample another day, at the cost of losing the site and redirecting us to the backup. They had calculated an 18 percent chance of waving off on approach to Nightingale, driven by the need to steer well clear of Mount Doom. Osprey was holding steady at 6 percent, making it the clear favorite of the engineers. Even though science was the goal, NASA had defined minimum success as simply contacting the surface.

I massaged my temples as the votes came in. Science and sampleability voted in favor of Nightingale. Even though I knew it was coming, I gasped as both the Lockheed and Goddard board members voted to make Osprey the primary site. The board was deadlocked. The decision was now up to me.

In their closing argument, Lockheed senior management stressed that Osprey was the first site to open up as we reduced the margins on the safety requirements, meaning it was clearly the safest of the four. Engineers love to hold on to their margin, and Osprey allowed them to keep a chunk of it in their back pockets. Additionally, the navigation calculations showed that Osprey provided the highest probability of contact on the first attempt, with the lowest probability of a wave-off. It was the best chance of achieving minimum mission success. The new NFT software and its recent Hazard Map upgrade were making progress, but they were still in development. Working under the mantra of not counting your chickens until they are hatched, they impressed upon me the fact that there was always a chance that some unexpected bug could delay the final implementation on OSIRIS-REx. Osprey was so benign, they felt confident flying there even without the Hazard Map on board.

"Damn you, Bennu," I cursed under my breath. "Can't you make anything easy for me?"

I asked Heather to adjourn the board. I needed time to think.

We were all but guaranteed to make it to the surface if I chose Osprey. However, what good was contact if we couldn't collect a sample? We would only have to try again at Nightingale anyway.

In the end, I went with my gut. My salivary glands started dripping again as I thought about the amazing sample that Nightingale could yield. Since the earliest days of our arrival, I had known instinctively that this was our site, almost as if it called to me. Even with Mount Doom hovering on the periphery, ready to take down OSIRIS-REx if it made one wrong move, I had to let the science guide me. When the board reconvened, I announced my decision—we were going to sample Nightingale.

INTERLUDE

CARBON CALLING

THE TERRESTRIAL CARBON ATOM JOINED a chain of nucleic acid, and with it came something significant in the history of life on Earth: the emergence of information. This information gained control over matter and energy, directing the flow of life on the planet, and bringing power, motivation, and purpose. Over time, the information grew in size and complexity, developing machinery to record, recall, reproduce, and iterate. Every new generation was an improvement on the last, diversifying and changing the planet in profound ways.

The carbon atom journeyed through the biosphere for billions of years, participating in biology's greatest inventions. It started with chemosynthesis, extracting energy and nutrients from the rocks. It invented nitrogenase to pull nutrients from the atmosphere. With photosynthesis, it harnessed the power of sunlight.

Sunlight provided a limitless energy source that fueled the growth of complexity and evolution. As plants emerged, they pulled carbon dioxide from the air, producing an unexpected byproduct: oxygen, a toxic and corrosive gas that made things *burn*. In response to this new environmental challenge, some organisms sought refuge inside other cells, a phenomenon known as endosymbiosis. Through this process, some cells morphed into a nucleus that became the repository of the genetic library. Other organisms evolved specialized cells for action, such as muscles for movement, and neurons for information processing. These neural cells gathered together, creating a mind with the ability to think, ponder, and marvel at the mysteries of the universe.

It was a journey filled with trial and tribulation, overcoming all through endless experimentation. The carbon atom lived a million lives and saw the birth and death of countless organisms. As life grew in complexity, the terrestrial twin found its purpose. It became folded into the genetic code of human beings, scientists and dreamers who quested for knowledge, and eventually a man named Dante Lauretta.

In the Arizona desert, the carbon atom compelled Dante to study the night sky, guiding him toward something out there.

"Come and get me," the wandering twin beckoned from within Bennu. "I have secrets to tell."

So, Dante did just that.

CHAPTER 13

TOUCHDOWN!

To remember the early days of March 2020 is to remember a different world. One in which our team, though scattered across the country, was reporting for work at office buildings, universities, and laboratories. We gathered in conference rooms and went out together for after-work drinks. Sure, we were stressed, we were tired, and we were staring down the daunting challenge of attempting to land a spacecraft on an asteroid—but we were also blissfully unaware of how much *more* difficult that was all about to become, thanks to a burgeoning global pandemic.

During the first days of spring, OSIRIS-REx set out on a daring mission, scouring Bennu's surface like a relentless raptor in search of prey. Venturing out of its Safe-Home orbit, the spacecraft targeted Nightingale, flying only 820 feet above the sample site. The objective was clear: Survey the site with stunning detail and uncover any hidden dangers that previous sweeps may have missed.

With its science instruments trained and ready, OSIRIS-REx squeezed every last bit of information from the site. The images were fed into the spacecraft's upgraded navigation system, creating a comprehensive Hazard Map that would guide it on its final descent. As it approached, OSIRIS-REx would compare what it saw with the Hazard Maps in its memory, ready to wave off at a moment's notice if any obstacles lay in its path.

The flyover was a triumph, and as OSIRIS-REx eased back into orbit, it changed direction, positioning itself for its next close encounter with Bennu: the first rehearsal for the ultimate goal, sample collection. The stakes were high, and the pressure was immense, but the hunter was ready, armed with the knowledge and the tools to pluck its treasure from the asteroid's surface.

The data from this flyover confirmed that Nightingale was still a "go." Things were going according to plan. We would perform the "CheckPoint" rehearsal in April, the "MatchPoint" rehearsal in June, and OSIRIS-REx would collect a sample of Bennu on August 25, 2020. When that happened, hundreds of us would gather in Colorado for the maneuver we had spent sixteen years developing—and the raucous party that would come after.

And then, COVID-19 wreaked havoc on our lives and our minds. Office closures, school shutdowns, and stay-at-home quarantine orders were issued. I left my house once a week for groceries, armed with latex gloves, two surgical masks, and tight, form-fitting clothes. A few weeks later, the lockdown was extended. Internet news sites became an obsession, filled with haunting images of teeming emergency rooms and eerily empty streets. The world, it seemed, stood completely still. After years immersed in the hurry and worry of space travel, the contrast was wild, dizzying, and dismaying.

It quickly became clear that this wasn't going to be a week- or even month-long crisis, but that it would stretch out at least though our rehearsal phase. All the stress and anxiety I had felt about sample collection was now compounded by a much bigger concern: keeping everyone safe while pulling off the mission in a barely functioning world. During those first weeks of lockdown, when society was descending into panic and uncertainty, there were moments when I honestly wasn't sure it would be possible.

Of course, I couldn't let myself think that. And I didn't have much time to think it, either. As March dragged into April, we were meeting multiple times a day virtually to discuss the lead-up to sample collection, and how to make it all happen given our now-limited capabilities. Because we had always been geographically diverse, we were all adept at online meetings, but now, screaming babies, stressed-out partners, and nosey dogs filled our screens; busy living rooms and poorly lit basements replaced the diploma-strewn offices I was used to seeing behind my colleagues.

My own home was also suddenly quite crowded. Though I was extremely lucky—Kate and I were parenting mostly self-sufficient tweens, and the four of us had room to spread out—the transition was rocky. After a few days of working from my perch in the living room, I just about lost it when my video conference froze up—again. As had become my habit, I stalked through the house, interrogating everyone's internet habits, seeking the bandwidth bandit. Each time it was a different culprit. Though Xander was in his virtual schoolhouse, he thought it was a good idea to simultaneously download a video game update on the PlayStation. Griffin became fixated on an iPad game with a seemingly limitless appetite for our data.

Once I got my terrestrial bandwidth situation worked out, we had bigger problems to solve. With the majority of us working remotely and often during irregular hours, the remaining effort to prepare for CheckPoint rehearsal was going to be an uphill battle. If I thought my internet connection was lousy, it was nothing compared to the trickle of data that OSIRIS-REx would be sending back to Earth during TAG. The range between the spacecraft and Earth had been increasing steadily, with it now on the other side of the Sun. As the distance increased, the data rate dropped proportionally. Due to this, and the fact that it was approaching an asteroid surface surrounded by giant boulders, including our nemesis Mount Doom, the spacecraft could only send back forty bits per second. That is "bits," as in a single "one" or "zero." A text message uses eight bits per character. Imagine reading your texts at a rate of five letters per second. It was almost enough to drive a person mad.

Because of COVID and our stressed-out work environment, we lost the efficiency that comes with having parties in the same room, and so it took longer to resolve issues that came up. Still, we continued working toward Check-Point rehearsal in April while recognizing that we couldn't anticipate all the challenges—coronavirus or otherwise—

that awaited us. If at any time we encountered a problem that put our personnel or our flight system at risk, we would immediately stand down, delay the maneuver, and leave the spacecraft in its Safe-Home orbit for the foreseeable future.

Even though we had decided to carry on, we couldn't control what happened outside of our bubble. One huge concern was the Deep Space Network, which was our direct line of communication between Earth and OSIRIS-REx. In early April, Spain was dealing with its first catastrophic COVID outbreak, and station management had reduced the Madrid facility to a skeleton crew to minimize the risk of infection. While the facilities in California and Australia were still up and running, the potential loss of the Madrid antennas, even for a few weeks, would seriously hamper our ability to talk to OSIRIS-REx, out there all alone at Bennu. We would lose valuable time responding to spacecraft alarms and navigation updates. In short, we would be flying with severely limited visibility.

Still, the weeks marched on, and the day of CheckPoint rehearsal arrived. On command, OSIRIS-REx sprang into action, executing the TAG sequence and firing its thrusters for the orbit-departure burn. As it trained its NavCam on the surface, it snapped pictures of the rugged terrain below. This time, something different happened. In the past, it had simply stored the data in its memory banks, waiting for the next opportunity to transmit the information back to Earth. After this sequence, however, it analyzed the images on its own, as if awakening to a new sense of self.

OSIRIS-REx recognized features on the surface and cor-related them with its extensive catalog of landmarks spread across Bennu. It used that knowledge to make a critical decision at the CheckPoint, four hundred feet above the surface. As the spacecraft calculated its position and velocity, it adjusted its trajectory to optimize the descent toward the surface. For the first time, OSIRIS-REx was truly autonomous, fully in charge of its own destiny.

For those of us back on Earth, the day of the CheckPoint rehearsal was a nerve-racking one, as a limited number of personnel monitored the spacecraft's telemetry while the rest of the team followed from their homes. In Tucson, a few of us gathered—masked up and distant—in a conference room at the Drake Building to witness the event.

After the spacecraft performed the CheckPoint maneuver, it descended for another nine minutes on a trajectory toward the asteroid. Upon reaching an altitude of approximately 210 feet, it performed a back-away burn to complete the rehearsal. We were tantalizingly close to Bennu.

During the rehearsal, the spacecraft successfully deployed TAGSAM from its folded, parked position out to its sample-collection configuration. Additionally, some of its instruments collected navigation images and made scientific observations of the site, providing a practice run for the incredible science that would be conducted during TAG. We were thrilled it had accurately estimated its position and speed relative to Bennu during the descent.

As the imaging sequence neared its end, Nightingale came into view at the bottom of the frame, with Mount Doom ominously perched on the southeastern edge of the crater's rim. The spacecraft had to fly right over the top of this nightmare before dropping down into the crater, ready to strike gold on the asteroid's surface. As the images came down of OSIRIS-REx approaching Nightingale's rocky floor, I almost wished we had scrapped the rehearsal and just touched down that day—that's how ready I felt we were.

I sat in the Drake Building conference room, relieved beyond measure to be back with my team, even for this brief moment. We were spread out, and our voices were muffled by masks. Nevertheless, we had done it; we were in the home stretch of sample collection. It was a bittersweet moment. Instead of bolstering our confidence and generating the unbridled enthusiasm of making spaceflight history, we just sank into our office chairs, exhausted and on the verge of a mental collapse.

While CheckPoint rehearsal was a rousing success, it had become increasingly obvious that folks were struggling. Heather, who had shown herself to be a deeply empathetic leader, had been hearing from teammates all over the country that they were tired, burnt-out, and juggling an inordinate number of responsibilities between the mission and their personal lives—both of which were now playing out entirely in the often-tight confines of their homes. She warned me that if we kept pushing forward at this pace, some of our team members may ultimately break down.

I decided to slow down, taking advantage of the extra time we had built into our schedule. We agreed to delay the final MatchPoint rehearsal until August and TAG in October. Maybe, I let myself believe, by October the pandemic would have receded enough for us to gather in Colorado and have that huge bash, after all.

The break gave us all time to rest and regroup. I sure needed a vacation. In true COVID form, that meant ordering a Nintendo Switch and spending two weeks playing Zelda with my boys.

✦ ✦ ✦

By mid-August 2020, the world was still nowhere close to back to normal. But the OSIRIS-REx team was ready to carry on. With the world in chaos, we became a beacon of hope for millions of people. Media interest spiked and I fielded countless invitations for interviews and presentations. With my living room in the background and the occasional appearance of my dog, I shared our adventure with the world. One of Kate's friends called her up and exclaimed, "I saw your living room on TV today!" No mention of her husband.

It was time for MatchPoint rehearsal, the final practice run before sample collection. This would be our last chance to make sure all of the instruments were ready for our *moment of terror*. MatchPoint was the instant when the spacecraft matched Bennu's rotation in order to fly in tandem with the asteroid's surface, 180 feet above the sample site, before touching down at Nightingale.

Just like the CheckPoint before it, our MatchPoint rehearsal executed perfectly. This final step before the big show allowed us to verify that all the spacecraft's systems were operating as expected. This rehearsal was also the first time that the Hazard Map was employed, the one we had created from all that high-resolution data. All signs back on Earth pointed to NFT being up to the task.

The team and the spacecraft were ready for sample collection come October. A certain weight slid off my shoulders after MatchPoint, not dissimilar to the feeling I had after our third proposal to NASA all the way back in 2011. We had done everything we possibly could to prepare for sample collection—and we had done it during an extraordinary global crisis. Whatever happened now was out of my hands. On October 20, 2020, we would find out whether the mission would be a success.

In early October, two weeks before TAG, I drove from Tucson to Denver. As the red desert gave way to yellow aspen, I was surprised by how calm and confident I felt. All the anxiety of the previous months had been replaced by an excited anticipation. Even the overwhelming heartbreak of the pandemic—and how it would prevent us from being together during this momentous occasion—had worn off a bit. If COVID had taught us anything, it was to roll with the punches, no matter how bad they stung. Still, I for one could not stay away from Lockheed Martin and the spacecraft operations. NASA was intent on a media blitz for TAG day and, once again, I was to be the on-air talent.

The subsequent days buzzed by, as the two dozen or so team members who had gathered in Colorado performed the final safety checks while senior management from NASA looked over their shoulders and signaled their approval. We also developed contingency plans for a wave-off, should OSIRIS-REx detect danger during its descent and ultimately decide to back away from Nightingale at the last minute. If that happened, the spacecraft's thrusters would disrupt the site's surface so significantly we would be forced to try again at Osprey. During the many months of rehearsals, we had prepped for that site as well, but the idea of additional weeks and maybe months of work felt overwhelming. Could our exhausted, overworked group rally from that kind of setback? I didn't want to find out.

The day before TAG, I turned fifty years old. I spent my

birthday in countless rehearsals for our big television show the next day. Heather filled my loaner office with black balloons and a giant inflated "50" Mylar reminder of my advanced age. Everyone gathered in the conference room to sing "Happy Birthday." My wish? A safe and successful TAG and a return to normalcy.

That night, alone in my Airbnb with the remnants of a store-bought cake stowed safely in the refrigerator, I took stock of the last two decades. When Mike recruited me for an asteroid sample-return mission, I was in my early thirties, new at marriage, parenting, professorship. Now, my kids were cruising toward high school and my beard was turning gray. Part of me still felt like that amateur sipping a whiskey at the Arizona Inn; in fact, part of me still felt like that little kid lost in the desert, looking up at the night sky, full of dreams and inexplicable desire. But mostly I felt battle-tested and ready.

✦ ✦ ✦

Our tiny team assembled in the Lockheed Martin mission control center, each of us wearing our blue NASA polos. Heather wouldn't have missed this event for the world, and despite her small stature, she exuded a calm and assertive presence. She was a natural in front of the camera, eagerly sharing her enthusiasm and expertise. Dani and Carl were back in Tucson, at the ready for the pictures to come down from the spacecraft the next day.

The room was divided into two sections—one where the engineers sat in front of monitors watching the continual trickle of information arriving from space, and another set up like a television set, with a Bennu backdrop and a host to interview a revolving cast of mission staff.

Broadcast began at 3:00 p.m. local time, and soon, more than a million people were tuned in to watch us make history. We had rehearsed the show a dozen times leading up to this event, but in every rehearsal, the timing was off. We never made it through the content before the big moment. I was a bundle of nerves, trying to keep track of the wall clock while listening to the mission controller read out the critical milestones as reported by OSIRIS-REx.

Out in space, OSIRIS-REx reached the CheckPoint and began its final descent, retracting its solar arrays for safety and approaching the surface at a steady pace of four inches per second. Armed with newfound knowledge, it expertly adjusted the MatchPoint maneuver to target the heart of

Dante and host on the set of the television broadcast—October 20, 2020. Credit: NASA.

Nightingale; bull's-eye TAG was now fully operational. With its smallest antenna in view of the Earth, it dribbled information back to the control room.

The eighteen-minute time warp that is communication across the solar system meant that every few minutes, a new data point would come down from OSIRIS-REx relaying the events of the past. I hungrily consumed every informational bread crumb. An hour in, we got the critical callout: "MatchPoint burn is complete."

This milestone meant that the spacecraft had fired its thrusters for the MatchPoint burn to slow its descent and sync up with the asteroid's rotation. It then continued a treacherous eleven-minute coast over Mount Doom, targeting a clear spot the size of a small parking lot in a crater on a pile of rubble that had been cruising through the solar system for hundreds of millions of years.

It's already happened, I thought to myself. Did OSIRIS-REx go for it? Or had the Hazard Map kicked in?

On the other side of the Sun, OSIRIS-REx hovered above its target. Its computer continued to process NavCam data, analyzing each pixel as the surface features became better resolved. With each snapshot, it ran calculations, weighing the odds of reaching a green zone or contacting a red location. Its fate and the critical decision to proceed or retreat was in its own hands.

As TAGSAM closed in on the surface, the moment of truth approached. At a range of just sixteen feet, OSIRIS-REx analyzed the final images, considering all its options, before making its decision. It then transmitted its choice back to Earth, where we were waiting to find out what had happened eighteen minutes in the past.

For the first time in my life, I realized the true enormity of the solar system, and it was vast beyond comprehension. Now my nerves were showing. I knew I was speaking rapidly and breathing heavily. So much for the rehearsals. It felt like the whole world was watching me freak out in real time. I would much rather have been over in the control room with my team, watching the telemetry pages, even though they were refreshing at a dismal five characters per second.

Three minutes later we heard the callout, "Attitude control system has transitioned to touch-and-go mode." Our knowledge about the final critical decision was only moments away.

"OREx is descending below twenty-five meters."

In the last few moments before contact, I reminded our host that the five-meter crossing is the crucial moment when OSIRIS-REx makes the go-no-go decision. "All my senses are on that callout right now," I told her.

The next callout came across the speaker.

"OREx has processed its next image, position uncertainty is 0.5 meters."

I could hear the team erupt in cheers.

"Hazard probability is 0 percent."

My heart started hammering in my chest. I knew at that moment that OSIRIS-REx had gone in. All that was left was for the data to make its way across the solar system.

"TAG lateral velocity is 0.2 millimeters per second."

"TAG vertical velocity is 10.2 centimeters per second."

All statistics were exactly perfect; the spacecraft was operating flawlessly.

The video feed switched over to the control room. I

stared at the screen, maintaining my focus on the team, anxiously awaiting the next callout.

My gaze focused on Estelle, who was monitoring the telemetry from Bennu. She would be the first to know that OSIRIS-REx had touched down. Everyone else's eyes were on her as well. Her spine straight, every once in a while she flicked her hands, as if shaking water from them, the only sign that she may be as nervous as the rest of us.

My heart was pounding. Sixteen years of work came down to these perilous few seconds.

After an eternity, the call finally came through.

"OREx has descended below the five-meter mark. Hazard Map is GO FOR TAG. Contact expected in fifty seconds."

I was done with the broadcast. I made one parting remark to the host, "We're going in!" and then I was out of there. Without another word, I squeezed past the Bennu backdrop of the soundstage and ran over to be with my team.

On my way over to the control room, I heard the critical callout.

"Touchdown declared."

I thought about the long, winding path to this moment, beginning with my lonely childhood spent staring up into a dark desert sky, Mike and the thousands of people who had worked across the planet to make this mission happen.

I thought about my family back in Tucson, watching me on television during the biggest moment of my career.

And of course, I thought about OSIRIS-REx, all alone on Bennu.

All at once, Estelle sprang from her chair, threw her gold-spangled wrists in the air in a manner that would make an NFL referee proud, and yelled the words I had waited for so long to hear:

"We have touchdown!"

Constrained by the health guidelines, we reached out our arms to each other and slapped virtual high fives, the pain of the separation muted only by the magnitude of our accomplishment.

A piece of primordial rock that had witnessed our solar system's long history may now be ready to return home for generations of scientific discovery, and I couldn't wait to see what came next. Of course, Bennu would have at least one more surprise in store for us.

CHAPTER 14

DEPARTURE

Many hours later, around 2:00 a.m., I leaned against the headboard in my rented apartment, laptop open, a cup of coffee on the nightstand. After TAG, I had endured a barrage of interviews, standing in place while a line of journalists waited their turn to get a sound bite. I was grateful for the media attention—and more than excited to brag about OSIRIS-REx—but that night, still shell-shocked from the successful TAG maneuver, all I could think about was what was coming back to Earth. The mission, after all, wasn't simply to touch the asteroid; we had to scoop up pieces of it too.

I wasn't the only one awake. The data were scheduled to arrive any minute and the science team chat rooms were abuzz with activity. Everyone was rehashing the best moments of the evening and eagerly awaiting the photos. The first clue that we had successfully collected a sample would come from comparing the images we had taken of the surface the moment before TAG and the picture captured one second after contact. If that second shot showed a significant disturbance of the surface, we probably had a decent-size sample on board. The key to successful sample collection was to get TAGSAM flush with the asteroid and hopefully buried a few inches. If any part of TAGSAM was lifted off the surface—say, tilted up due to a rock under its lip—the gas would blow out to the side, sending our precious sample flying up and away.

Right on schedule, the data started arriving. The spacecraft had taken eighty-two images during its descent. We expected to see TAG around frame seventy. About one scene per minute posted on the science data server. That meant, tucked in my rented bed, I would have to wait an entire agonizing hour to see how OSIRIS-REx interacted with the surface. The minutes crawled by.

Then, the last image before TAG came down, and, one minute later, so did the first post-contact shot. As soon as I saw that view, I knew we had the sample. With this realization, a wave of emotions swept over me—overwhelming joy, relief, and a profound sense of fulfillment. After all the challenges and hardships we had faced, I almost couldn't believe it. We had actually done it.

I set the two images blinking back and forth, watching with rapt fascination as TAGSAM hit the surface of Bennu then sank right into what appeared to be a spongy asteroid surface. Everywhere I looked there was evidence of disturbance. A large rock right under the sample head appeared to completely disintegrate on contact. Another rock, roughly twice the size, responded like a seesaw, tilting and launching small particles into space.

Like a well-placed arrow, OSIRIS-REx aimed true and the TAGSAM arm had contacted Nightingale. After ten minutes or so futzing with the pre- and post-contact images, several more shots were now on the server. As I downloaded them, I was stunned to see Bennu's surface erupt in a shower of particles, many captured in the waiting sample collector when it unleashed its surge of pressurized nitrogen gas. I watched in rapt fascination as the TAGSAM head descended into shadow. Five seconds later, the spacecraft's thrusters roared to life, applying the brakes and stopping its plunge into Bennu, kicking up a cloud of loose rock in its wake.

Our simulations had suggested that the head would penetrate a fraction of an inch into the subsurface. The shadowy images showed that we had gone much deeper than that. The arm had penetrated deep, sinking almost two feet into the surprisingly delicate crust, a testament to its remarkable fragility. After sixteen seconds, the arm emerged triumphant, clutching a precious haul of Bennu's riches. Soon, OSIRIS-REx's thrusters fell silent, and it drifted away from the asteroid, a daring adventurer, ready to share its incredible findings with the world.

As more data came down, the true nature of the surface disturbance became apparent. The spacecraft's thrusters had mobilized a substantial amount of surface material during the back-away burn; a massive plume of loose debris was visible near the end of the sequence. Nightingale was unrecognizable.

In addition to these encouraging photos, the spacecraft reported back that it was still in fighting shape. Despite the barrage of dust and rocks it had kicked up, OSIRIS-REx had not suffered any noticeable damage. We'd worried that grains would clog the engines, or that dust on the radiators would prevent thermal energy from leaving, causing the electronics to overheat. But all thrusters were firing, and the onboard radiators were keeping the spacecraft cool. The solar arrays and star-tracking navigation system, despite likely picking up some dust from the impact, were working perfectly.

The spacecraft knew where it was in space and appeared to be ready and able to fly home. Once we knew how much sample we had on board, we would direct it to do just that.

Over the next few days, we would determine the *amount* of sample collected. We would scour pictures of the TAG-SAM head itself taken by SamCam, the spacecraft's camera devoted to documenting sample collection, to ascertain whether dust and rocks made it into the collector head. We would also peer inside of the head so we could check for evidence of sample.

But our strategy involved much more than just visual confirmation alone. Because stowing the sample in the sample-return capsule meant severing the collector completely from the robotic arm, we needed to be absolutely positive we had successfully collected bits of Bennu. If not, we would attempt another TAG maneuver a couple months later at Osprey. Therefore, the mass of our sample would need to be measured from two hundred million miles away.

Dante describing the TAGSAM contact with Bennu on October 21, 2020. Credit: NASA.

We aimed to accomplish this by calculating the change in the spacecraft's "moment of inertia," a term that describes how mass is distributed and how it affects the rotation of a body. This "spin test" maneuver was executed by extending the TAGSAM arm out to the side of OSIRIS-REx and gradually rotating the spacecraft. A motion reminiscent of a person spinning with an arm outstretched, holding on to a string attached to a ball. As the person spins faster, they can gauge the ball's weight by the tension in the string. Similarly, we would measure the mass of the collected sample by monitoring the energy required to spin the spacecraft. By performing the spin test both before and after TAG, we would be able to calculate the change in the collection head's mass, indicating the presence of the gathered sample.

Between the images and the mass measurement, we hoped to confirm that we had collected at least those two ounces. If our confidence was high, I would make the decision to stow the sample on October 30. Bennu, ever the trickster, had other plans.

Two days later, I stood next to Estelle in the spacecraft control center at Lockheed Martin, once again waiting for images. A few hours earlier, the spacecraft had executed a veritable TAGSAM photo shoot, taking pictures as the robotic arm moved through several different positions. At each of nine positions, the camera snapped a range of exposure times from milliseconds to seconds. I looked around at the small contingent of engineers, their focus all trained on that screen. I couldn't help but smile as I remembered the many times we had done this now, gathering and waiting on a crucial image from space. We were, I realized, running out of these moments as the mission hurtled toward the last lap.

The first shot was a short exposure with TAGSAM positioned ninety degrees to SamCam—the perfect profile pic. The entire operations team was gazing at the big screen when someone piped up, "Hey, I see a particle!" In unison, we all took a step closer to the screen to see what he was pointing at. Sure enough, it looked like a dust grain was floating in the vicinity of TAGSAM.

"Don't get too worked up," came a voice from the star-tracker engineer. "We've seen the spacecraft shedding a bunch of dust ever since we backed away. We definitely picked up some hitchhikers—seems they are heading out on their own now."

As the images continued to come down, we noticed a streak or two in the frames. At this point, we were experts in streaking particles. And now we had the tools to track these escapees back to their source.

In the longer exposure photos, there were even more long streaks emanating from TAGSAM. The streaks continued through each exposure and each position. As we animated the entire sequence, TAGSAM reminded me of one of those lawn sprinklers, the ones that spray an arc of water from one side to the other and then cycle back to start again.

Estelle and I turned to one another at the same time. "I don't like this," I said flatly.

"I don't like it either," she agreed, worry creeping into her eyes. "What the heck is going on?"

I stepped out into the hallway and dialed the person with the most experience in analyzing such data—Carl.

He picked up the phone immediately. "Hey, Dante, what's up?"

I imagined him running through his own mental checklist, wondering what could have possibly prompted me to call him at this time.

"Have you seen the latest SamCam images?" I queried.

"No," he replied. "What's going on?"

"I think we are seeing some particles escaping from TAGSAM. Can you take a quick look?"

An answer came quickly: The particles definitely originated from TAGSAM. These weren't hitchhikers on the outside of the spacecraft, they were coming from *inside* our sample head itself.

I headed back into the control room to give Estelle the grim news.

"Looks like we are spewing sample into space," I told her.

As I flipped through the animation of TAGSAM's motion, I could start to trace individual particles as they flew away.

"We're bleeding," I told her. "We've got to find a way to stop these particles from escaping."

Carl estimated that we were losing ounces of material in just the few moments caught on camera. The situation had turned into an emergency. Clearly, Bennu was not done messing with our minds just yet. As I sat helpless in Denver, OSIRIS-REx was out in space, spilling the sample we had spent decades and nearly a billion dollars to retrieve.

As we pored over the SamCam images trying to diagnose the problem, it became clear that TAGSAM was indeed full of rocks and dust. We could see bits of material passing through small gaps where the head's mylar flap was wedged open. The unsealed areas appeared to be caused by larger rocks that didn't fully pass through the flap. Some of the escaping particles were bigger than half an inch across. They were also flaky. The scene looked like someone had dumped a box of corn flakes into space.

I thought to myself, *Each one of those particles could have been someone's PhD thesis.*

If we stuck with our original plan, we would be executing our mass measurement maneuver. But accelerating the spacecraft around in a circle would almost certainly exacerbate the leakage of the sample. OSIRIS-REx would look like a fairy out there in the dark, sprinkling glitter all over the cosmos.

I found Estelle seated at her console. "How fast can you stow the sample?" I asked.

She didn't hesitate. "We can start this weekend if we get approval from NASA."

I nodded, then called a halt to all spacecraft activities that could induce vibrations or particle motion. From here on out, OSIRIS-REx would focus solely on stowing the sample in the sample-return capsule, where any loose material would be kept safe during the spacecraft's journey back to Earth.

I got on the phone with NASA Headquarters and explained the situation. With the spacecraft in quiet mode, the number of particle streaks crossing the star-tracker field of view had dropped back down to zero. We had managed to stop the bleeding.

By the end of the afternoon, based on available imagery, we suspected there was plentiful sample inside the head, a fact that otherwise would have been fantastic. Our best guess, based on the particles we could see both jamming open the flap and inside the now-exposed collection chamber, was about fourteen ounces, more than seven times what we had been aiming for.

Ultimately, the sheer abundance of material that appeared to be in the sample head made it possible to expedite the decision to stow. But now we would have to work nonstop to accelerate the maneuver we had assumed would take weeks to implement—and do it while protecting as much of the sample as possible.

Unlike other spacecraft operations, during which OSIRIS-REx autonomously ran through an entire maneuver sequence, stowing the sample was done in stages. Each step required precision geometry, to ensure that the sample head locked into place inside the return capsule. The mechanism looked a lot like the interface between ski boot and ski. The last thing you wanted to do was head downhill without your boot clicked firmly in place.

Thus, stowing the TAGSAM head required constant oversight and input. We would send the preliminary commands to the spacecraft to start the stow sequence. Once OSIRIS-REx completed each step, it sent images back to those of us on Earth and waited for confirmation before proceeding with the next one. It was a true telerobotic operation, requiring a painstaking forty-minute span between each radiated command and return signal.

The approval to expedite stow came quickly from Headquarters. They had watched this team perform flawlessly under extreme conditions—the head of NASA science even compared us to Olympic athletes—and I looked on with enormous pride as they once again sprang into action. Over the past four years, this group had bonded, and we now operated as a unified whole. Everyone knew their role and the stakes couldn't be clearer. Mission success was on the

Particles escaping from TAGSAM. Credit: NASA/Goddard/ Lockheed Martin/University of Arizona.

line. With their orders in place, they set about to save as much sample as possible and get TAGSAM safely tucked away into its protective return capsule.

A few days later, I stood in the shadows of the space-craft control room and stared in quiet confidence as the first image from StowCam (yet another onboard camera built to, as the name implies, document sample stowage) showed the collector head hovering over the capsule after the TAG-SAM arm had moved it into the position for capture. The next image showed the TAGSAM head secured onto its custom-fit capture ring. I experienced another pang of anxiety as a small cloud of particles emanated from the head as it clicked into place.

Hopefully, I pleaded internally, *those are the last particles to escape our grasp.*

After the head was seated into the capsule's capture ring, the robotic arm performed a safety check. This maneuver tugged on the collector head and ensured that its latches were well secured. Following the test, we watched the sequence that showed TAGSAM tucked in for its long journey home.

Before the sample head could be sealed into the capsule, two mechanical parts on the TAGSAM arm had to be disconnected—the tube that delivered the nitrogen gas and the bolts that connected the head to the robotic arm. Over the next several hours, the engineers commanded the spacecraft to cut the tube and separate the bolts. We were looking at a decapitated TAGSAM. The head was locked securely in place and tubes and wires dangled off the top, looking like hair that had been cut with a cleaver.

That evening, the spacecraft completed the final step of

sample stowage. To secure the capsule, the spacecraft closed the lid and then fastened two internal latches. Our baby, as the ubiquitous bumper stickers announce, was on board.

Not deterred by the loss of the sample mass measurement opportunity, the engineers came up with a new way to estimate the mass of the collected sample. They were able to track the small forces on the spacecraft as we swung the TAGSAM arm through its various photo-shoot positions and the final stow procedure. Analysis of these minute changes indicated that at two days post-TAG, TAGSAM contained over ten and a half ounces of sample. By the next measurement, eight days post-TAG, just before sample stowage, the same technique was used to measure a sample mass of just under nine ounces. I'll admit, it was a bit of a gut punch to realize that a sample almost the size of the entire mission requirement was lost in space.

Still, as best I could tell, we had accomplished everything we had set out to do. We had touched the asteroid and garnered what appeared to be a massive sample. We had then successfully stowed it away, locked up tight in the heart of the return capsule, buckled in for the long journey back to Earth. Indeed, it was time to call OSIRIS-REx home.

Except, one thing continued to nag at me.

After seeing the dynamic response of the surface, I was bewildered by the abundance of boulders and pebbles strewn about, given how gently the spacecraft tapped the asteroid. Bennu's surface was considerably disturbed by the sample-collection event.

But every time we had tested TAGSAM in a lab—or even on board the Vomit Comet—we had barely made a tiny divot. The surface response was so different from our predictions, I knew we were missing some fundamental knowledge about how these tiny rubble-pile asteroids behave. The scientist in me could not live with the fact that we would leave Bennu with a mystery hanging over our heads. We had to go back and take one final look at the mess we had made.

The problem was that Lockheed's contract explicitly stated that there would be no science measurements after TAG. The reason for this overly specific legal clause was lost in the annals of history. However, I was not going to let some bonehead lawyers stand in the way of my science. I decided to make my case to senior management at Lockheed, to see if we could make one last detour.

Ultimately, it was an easy sell. The spacecraft showed absolutely no sign of any degradation in performance. I wanted to go swooshing in like those very last passes we had made prior to TAG, documenting the site at the highest resolution feasible. But those trajectories were not without risk, including possible collisions with Bennu.

Instead, the mission team quickly completed a detailed safety analysis of a new trajectory to observe Bennu on our way out. This flight path would keep OSIRIS-REx at a safe distance from the asteroid, while ensuring the science instruments could collect the needed information. Just like the early days, we would scan the whole surface, then focus on the few images that covered our region of interest. It was bittersweet to plan one more maneuver at Bennu—not only to provide even more information but to also bid a proper goodbye.

In April 2021, OSIRIS-REx returned to the fray and

cast its gaze one last time upon the enigmatic asteroid. The spacecraft dove in for a close encounter, flying within 2.3 miles of Bennu's surface — its closest approach since the historic sample-collection event half a year earlier. For almost six hours, the spacecraft monitored Bennu, collecting data and snapping images as the asteroid spun below.

Just as we had hoped, these new images revealed the aftermath of the astonishing encounter with Nightingale. The comparison of the new data with previous high-resolution shots of Bennu from 2019 revealed striking signs of surface disruption. At the sample-collection site, a depression was visible, with several boulders noticeable at the bottom, indicating that they were uncovered during sampling. The area near the TAG point was now more highly reflective and had an increased concentration of bright material, almost like a sprinkle of salt against the dark background of the surface. Despite being located within a dark red crater, the Nightingale site appeared to now have a salt-and-pepper texture. The sampling event left a stunning impact crater, much deeper than expected and measuring an impressive twenty-six feet across. I couldn't help but be filled with awe as I observed the indelible imprint we had made on the surface, reminiscent of the iconic record symbolized by Neil Armstrong's boot print.

After analyzing the data, we discovered something remarkable. If OSIRIS-REx hadn't immediately fired its thrusters to back away after collecting its sample, it would have sunk into Bennu. The particles making up Bennu's exterior are so loosely packed and weakly bound to each other that if a person were to step onto Bennu, they would experience very little resistance, as if they were stepping into a ball pit at a children's play area. Until the very end, Bennu remained unpredictable.

On May 10, 2021, OSIRIS-REx roared back to life, firing up its main engines for the first time in three years. With a burst of power, it ran the engines at full throttle for seven minutes, propelling the spacecraft away from Bennu at breakneck speed of 600 miles per hour. As the asteroid shrank into the distance, OSIRIS-REx embarked on its two-and-a-half-year journey back to Earth, its precious cargo firmly in its grasp.

That evening, as the sun set over the jagged Tucson mountains, I went on a hike to clear my mind. When I reached the summit, the horizon was on fire with the reds, golds, and oranges of a gorgeous Arizona sunset. I lay back on my favorite boulder and studied the patch of sky where OSIRIS-REx was inbound. As I searched the heavens, a lone coyote howled in the distance, its mournful cry piercing the stillness of the desert. I could feel the weight of the past decade's work slowly draining into the Earth beneath me.

An overwhelming sense of calm came over me as I settled into the feeling that everything we had worked for was finally coming to fruition. As the cool evening breeze caressed my face, my thoughts turned to the pledge I had made years ago, the bold promise that underpinned our entire proposal to NASA, a vow to unlock the mysteries of the origin of life itself. We had come so far, but there were still challenges ahead. And yet, as I looked out at the endless expanse of the universe, my resolve grew stronger. We were on the verge of something truly incredible, and I felt in my soul that the final phase of our mission, sample analysis, would reveal the deepest secrets of the cosmos.

EPILOGUE—PART 1

HOMECOMING

On September 24, 2023, I awoke at 1:30 a.m. in my room at the Holiday Inn, a restless night behind me. Outside, Dugway Proving Ground, a military post in the middle of the Utah desert, was quiet. My health tracker told me my heart was pulsing at 120 bpm—a telltale 60 bpm above my calm baseline. Anticipation tinged the air.

I logged on to a video call with my team for the "go, no-go" poll, the final survey before our critical event. It was textbook OSIRIS-REx: flawless. Despite a few recent glitches in the spacecraft's gyros, all systems were nominal. Cheers erupted when the sample return capsule release was green-lit.

Now I needed distraction. With two hours to go, I immersed myself in grading essays for my astrobiology course, momentarily lost in the minds of my students. But the day's gravity kept pulling me back. The capsule's batteries had to depassivate, a crucial milestone. They had lain dormant for the past seven years. Without their trickle of power, the parachute would not deploy, and we would be coming in hard. As the spacecraft approached Earth, the tension among the team members was palpable. I watched my computer screen with bated breath as the batteries inside the capsule came online. Once the voltage reached 11.44, elation washed over me. We were live.

The spacecraft then executed a seamless capsule separation. My eyes clung to the telemetry and Doppler data. Our capsule, set free one-third of the distance to the Moon, began its four-hour plunge toward Earth.

Caffeinated and geared up in my freshly purchased field attire, I found myself in the lobby, where high fives with Mike Moreau, our deputy project manager, signaled the mood. Everything had gone as planned so far, but little did I know that the most heart-stopping moments were yet to come.

A few hours later, I was in the air, rotor blades chopping through the Utah sky, our helicopter slicing through the atmosphere in pursuit of a different kind of skyfall — our asteroid sample. The headset buzzed with chatter. The voice of the Air Force range command officer, a comforting presence over months of training, narrated the capsule's high-speed journey home.

Everything hinged on one pivotal element — the parachute deployment system.

The capsule's reentry into Earth's atmosphere was particularly intense. It charged in at a breathtaking speed of 27,650 miles per hour, creating a fireball as the capsule's momentous velocity translated into extreme heat. Seated in the front of the helicopter, the Air Force safety officer relayed the first hopeful news: the plasma trail had been spotted over California from the WB-57, NASA's high-altitude aircraft tracking our precious cargo.

It had hit the top of the atmosphere at incredible speeds, but I knew that it was designed for exactly this purpose. Even so, it was a moment of great tension. I imagined our capsule out there, hurtling through the air as the calls came through my headset. At entry plus 52 seconds, the capsule experienced peak heating. This was a tense moment, as the heat shield had to withstand extreme temperatures to protect the sample inside. With the searing heat, infrared tracking became possible, allowing our Air Force colleagues to monitor its descent.

Just ten seconds after peak heating, the capsule experienced a deceleration of over 30 gs, slowing down from its incredible velocity. I knew it was crafted to handle these extreme conditions, but I held my breath anyway. At entry plus 116 seconds, the deceleration decreased to 3 gs, and a timer through a g-switch was supposed to initiate the deployment of a drogue parachute, stabilizing the capsule for the next phase of its atmospheric passage.

My ears strained to hear the status calls. The capsule was at a hundred thousand feet, the critical altitude for the drogue chute to deploy.

"Any sign of the drogue?" I inquired, my voice tinged with urgency.

"They're not calling drogue," the safety officer replied.

The helicopter pilot gasped into his microphone as he heard the capsule had entered Utah Test and Training Range airspace at a jaw-dropping speed of 8,000 knots. I leaned over to Scott Sandford, my co-investigator, our eyes locking.

"This isn't good," I murmured.

Flashes of the 2004 Genesis disaster flooded my mind. We had labeled that calamity our worst-case scenario throughout countless design reviews and tests. Compared to Genesis, our payload was even more vulnerable — soft clay minerals, easily dispersed by the desert winds, a powder keg of scientific potential about to be lost.

My heart pounded in my chest like a war drum. My hands trembled. I thought of the cameras that would immortalize our failure, broadcasting it to the world. Desperation clawed at my insides.

And then, breaking through the static like a celestial anthem, the safety officer's voice: "Main chute has been spotted!"

The capsule was overhead, and it began its vertical descent. I couldn't contain myself. A triumphant roar erupted from me, startling the helicopter crew. Tears filled my eyes, blurring my vision, yet magnifying the sense of relief washing over me.

Hill Air Force Base tracked the capsule's descent using radar and infrared technology, providing us with precision knowledge of the landing site. Multiple tracking stations communicated the capsule's location to one another, supposedly maintaining a landed position accuracy better than thirty feet. However, confusion over coordinates led us on a merry chase. Our seeming inability to spot the capsule made my doubt linger. The absence of the drogue chute at high altitude was still a mystery; could it have somehow resulted in a crash landing after all?

Finally, we found it—our interstellar treasure, incongruously nestled just seventy feet from a road, as if placed there by an invisible hand. It looked unreal, like a prop from one of our training exercises. But the charred exterior told a different story, proof of its fiery journey through the atmosphere, temperatures soaring to 5,000 degrees.

And it was there, intact, unmoving, perfectly aligned like a gymnast sticking a landing in the Olympics. The capsule had created its own crater, its shield leaving an impression in the Utah soil, a mark not just on the Earth but on human understanding of the cosmos.

Dante and crew documenting the environmental conditions of the capsule landing site. Credit: NASA.

I watched from nearby as Lockheed's elite technicians meticulously secured the capsule in a specialized cargo net, hanging it from a hundred-foot tether beneath Helicopter One. I paused, captivated by the cinematic ascent as the helicopter gracefully lifted the netted capsule into the sky. It was as though the capsule, an emissary from the cosmic frontier, was now being guided by earthly chariots.

In pursuit, Helicopter Two took to the air, its cameras capturing every frame for posterity and television audiences. Not far behind, Helicopter Three, carrying Lockheed technicians, dusted off the desert floor, its rotor wash blurring the fine grains beneath. Their part in this grand orchestration complete, they too ascended, leaving us—a trio from the field team—alone on terra firma.

As the chorus of engines dimmed, swallowed by the vast Utah desert, the soundscape turned minimalist: just the whisper of a gentle breeze skating over an ocean of sand and stone. A wave of profound accomplishment swept over me.

In that stillness, my mind wandered to another isolated corner of Earth—Antarctica. I recalled the C-130 taking off, its roar dissipating into the icy silence, leaving us alone at the onset of a different kind of expedition. It was as if the Utah desert and the Antarctic wilderness were speaking the same universal language of solitude and exploration.

I allowed myself a moment to absorb the raw, unblemished beauty of the landscape—an unexpectedly welcoming haven for our cosmic guest. Being part of the team that ushered these ancient samples back to Earth and standing on the exact soil where they first touched home defied words.

In the vastness of the Utah desert, my thoughts floated skyward, latching on to our spacecraft still zipping past the

Earth. Even as we celebrated one mission's conclusion, another was taking shape. The spacecraft had just executed a deflection maneuver, positioning itself into a solar orbit to ensure it wouldn't be drawn into Earth's atmosphere alongside the capsule. NASA had rebranded our mission OSIRIS-APEX, charting a new trajectory toward another celestial body: Apophis—an asteroid teeming with scientific promise and public fascination.

While our previous mission to Bennu had been focused on sample collection, OSIRIS-APEX presented a new set of objectives. This wasn't merely a data-gathering expedition; it was a mission pivoted toward planetary defense, aimed at closely monitoring Apophis's orbit, composition, and interaction with Earth's gravitational field during its anticipated historic close approach in 2029.

Transitioning from studying Bennu, a B-type asteroid, to Apophis, a stony S-type asteroid, provided an unprecedented scientific opportunity—a comparative study of two vastly different cosmic entities. As we stood on the precipice of this new adventure, our team underwent transformative changes as well, embodying the mission's own evolutionary nature.

One of the most salient shifts came in the form of leadership. While I continued my role as the principal investigator for the OSIRIS-REx sample analysis campaign, the reins for OSIRIS-APEX were handed to Dani. Her ascent from undergraduate student to the role of principal investigator was emblematic of the limitless horizons in space exploration. Her journey, just like mine, was a resounding testament to the potency of education and mentorship—dynamic catalysts that drive both innovation and personal growth.

More than two hundred undergraduate and graduate students have contributed to this grand endeavor. Their youthful exuberance fueled our mission, reaffirming the importance of training the next cadre of cosmic explorers. More than a mission, OSIRIS-APEX served as a platform for continuous learning and scientific continuity.

A wave of immense pride washed over me as I pondered this shift. The transition from OSIRIS-REx to OSIRIS-APEX was not just a leap in mission objectives; it encapsulated Mike Drake's belief that through nurturing the next generation, we distill the wisdom of ages into groundbreaking discoveries.

So, as we face the vast, uncharted expanses of space, I'm reminded that our mission—our journey—is not just scientific. It's a deeply emotional and human endeavor. We are not just seeking to understand the universe; we are paving the way for future explorers to ask bigger questions, seek bolder answers, and touch the cosmic beyond.

As the helicopter descended onto the tarmac at Michael Army Airfield, I could sense the collective excitement awaiting us. A tent had been erected, hosting distinguished guests like the president of the University of Arizona alongside representatives from Lockheed Martin, NASA, JAXA, the Canadian Space Agency, and countless other well-wishers. My family—Kate, Xander, and Griffin—were visibly elated, their faces beaming.

Stepping out of the helicopter, I couldn't help but raise my fist triumphantly. The crowd erupted in cheers, palpable

Arrival at Michael Army Airfield after capsule recovery. Credit: NASA.

energy reverberating through the airfield. A hero's welcome awaited me as I made my way to the tent, where I embraced Kate and our boys, overwhelmed by a mixture of joy, relief, and the profound significance of the moment.

As I spoke to well-wishers and the media, in truth, my attention had pivoted to the status of our capsule, which was currently being examined in the clean room. Viewing the footage, it was clear the capsule had experienced anomalies during its descent when the drogue chute had failed to deploy at the high altitude as planned and the capsule had tumbled through the upper atmosphere. Yet, miraculously, the main chute had deployed during a moment of stability, enabling that pinpoint landing. I couldn't help but attribute this last-minute crisis and save to Bennu itself—the mischievous asteroid seemed to have one more surprise for us before yielding its secrets.

Later, I drove the few miles over to our temporary clean room, where meticulous work was already underway. The curation team operated with military precision, carefully placing the science canister into a protective Teflon bag, preparing it for its journey to Johnson Space Center. A C-17 aircraft, reminiscent of the one that had initially transported our spacecraft to Kennedy Space Center over seven years ago, awaited us on the tarmac. We were ready for the final leg of this chapter—delivering our hard-won samples to their new home for analysis.

The following morning was a mix of jubilation and anticipation. The hangar was a flurry of activity, resembling more of a festive gathering than a typical pre-flight preparation. Team members milled around, grabbing sack breakfasts and ensuring that the capsule was securely stowed for its onward journey to Houston. This invaluable material had survived an odyssey through space and now symbolized the epitome of human endeavor and collaboration.

The C-17 landed smoothly at Ellington Field, Houston, at 11:40 a.m., concluding the final part of a journey that had spanned billions of miles and countless hours of labor and anticipation. The atmosphere was electric; our team, alongside NASA staff, immediately transitioned to the specialized clean room at the Johnson Space Center, custom-built for Bennu's samples.

Being in that room felt like standing on hallowed ground. The glove boxes, meticulously designed, held within them the science canister—our window into the far reaches of the solar system. As we prepared to open it, I was reminded that we weren't merely examining rocks and dust; we were opening a treasure trove of cosmic history.

A collective gasp escaped our lips as the outer lid was lifted and TAGSAM was exposed. Inside, dark powder and grains of dust glimmered, as if knowingly holding on to their ancient secrets. For the first time in three years, this cosmic vault was unsealed, revealing hints of stories written billions of years ago, far from Earth.

Each grain we examined was a universe of its own, promising to illuminate the very questions that have captivated human curiosity for generations—how our solar system was born, what roles carbon-rich asteroids like Bennu may have played in seeding Earth with life's precursors, and perhaps even clues about how common life is across the universe.

The procedures following the opening were executed with surgical precision. Every particle was cataloged, every data point analyzed, but amid this rigorous science, there was room for wonder, and there was room for dreams—dreams of what these particles could tell us and about their impact on the generations of cosmochemists and astrobiologists who will take up the mantle after us.

Even as we reveled in this momentous occasion, we knew that the journey was far from over. This was just the first chapter in an extensive narrative that would span decades of research and exploration. It's a narrative that would be carried on by those we educate and inspire, who will continue to seek the extraordinary in the infinitesimally small, finding the universe in a grain of asteroid dust.

Later that evening, I sat back to ponder the odyssey that was the OSIRIS-REx mission. This mission, shaped by a collective dream of exploration and discovery, had an audacious aim: to pry open the secrets of our solar system by visiting Bennu, a near-Earth asteroid. This endeavor was not just a tribute to human innovation, but also to our relentless quest for understanding the cosmos. The mission promised revolutionary insights into a carbon-rich asteroid, a primordial relic that could potentially transform our understanding of planetary genesis and even the origins of life.

Among the numerous lessons, paramount were persistence and resilience. Space exploration is a perilous venture, a dance with the unpredictable and the unknown. I had affectionately termed Bennu the "trickster asteroid," and it lived up to that moniker, presenting us with enigmas and obstacles at every turn. From its topography—a conglomerate of rock, gravel, and boulders barely clinging together through gravity—to the unexpectedly fluid-like behavior of its surface during TAGSAM's contact, Bennu kept us on our toes. Yet adversity only steeled our resolve. We didn't just survive; we thrived, adapting and innovating to not just meet but exceed our goals. This fortitude, this innate human grit, has been a constant ally throughout my career, affirming that the quest for enlightenment is worth every tribulation.

Another cardinal lesson has been the profound impact of collaboration. I've stood shoulder to shoulder with some of the world's most brilliant minds, realizing that our most awe-inspiring accomplishments are born from unity and shared fervor. The abysses of space appear less impenetrable when we pool our resources and talents toward a mutual aim.

Leading OSIRIS-REx has been as much an honor as it has been a tempest of emotions. From the euphoria of the launch to the nail-biting moments of sample collection and, finally, the serene relief accompanying the capsule's safe Earthbound descent, I was continually reminded of the magnificent ensemble of scientists, engineers, and cosmic aficionados who had rallied together to manifest this vision. Their unyielding dedication and expertise formed the backbone of our triumph, their zeal for discovery utterly contagious. A profound gratitude washed over me; I felt grateful for the chance to be a part of this extraordinary cosmic pilgrimage.

My sojourn as an asteroid hunter has been deeply transformative, a voyage of internal and cosmic exploration. The quest was not just about reaching celestial landmarks, but also about plumbing the depths of my own psyche to distill the quintessence of what it means to be an explorer. Exploration, I've learned, is as much about the inward journey as the outward one, providing answers to age-old queries while generating new questions that propel us even further. This pursuit, both personal and collective, transcends the confines of time and space.

In essence, the OSIRIS-REx mission was a journey that stretched our technological limits and laid bare the indomitable human spirit, our capacity for ingenuity and steadfastness. I have been privileged to be not just a spectator but an active participant in this incredible epic.

My vision for the legacy of OSIRIS-REx is one of enduring inspiration and the perpetuation of human curiosity, ambition, and exploration. As we look to the future, our legacy will serve as a guiding light for generations of space missions yet to come. It will stand as a symbol of what humanity can achieve when we dare to dream, when we push the boundaries of our understanding, and when we reach out to touch the cosmos.

Ultimately, OSIRIS-REx is a call to action—an invitation for humanity to continue reaching for the stars, to embrace the unknown, and to discover the extraordinary within the ordinary. It is a reminder that the universe is our classroom, our laboratory, and our muse, and that our journey of exploration has only just begun.

EPILOGUE—PART 2

CARBON UNITED

THE SPACECRAFT EXECUTED ITS TOUCH-AND-GO sampling maneuver, stirring the surface and collecting pieces of ancient rock—among them, the entangled carbon twin.

When the sample returned to Earth, it was ultimately delivered to a laboratory in Arizona. The sample was processed, sieved, analyzed—and then it happened. An isotopic analysis of the carbonate minerals revealed an intriguing signature, a message from the cosmos that seemed to echo across eons.

The terrestrial carbon atom, ensconced in the DNA of Dante, felt a shudder of recognition. It was as if a thread of spacetime had been pulled tight, bridging an unfathomable distance. The carbon twins, reunited after billions of years, communicated in a language beyond human comprehension.

The carbonate mineral, bearing the wandering twin, was placed in a state-of-the-art mass spectrometer, its molecules vaporized and charged, ready for analysis. The data began streaming onto Dante's screen, forming peaks and valleys that hinted at the atom's origin. The information it contained was a Rosetta stone for astrobiology, a key to unlock new scientific horizons. It seemed as if the wandering twin whispered secrets from the cosmos into Dante's ear—clues about the solar system's early history, the conditions that led to the formation of life, and perhaps even the link between life and consciousness.

This was a moment of scientific triumph, but for Dante, it felt profoundly personal. This was not just about rocks and atoms; it was a meeting of kindred spirits. The terrestrial twin, once a mere carbon atom in the crust of a primitive Earth, had traversed the journey of life to become part of a sentient being capable of love, wisdom, and scientific inquiry. The wandering twin, ejected into the cold void of space, had found its way back to a world teeming with life, ideas, and possibilities.

And so the carbon twins were united, each contributing to a monumental leap in human understanding. The implications were staggering. As Dante looked ahead, he knew that the research would be groundbreaking, with questions leading to answers and those answers leading to more questions—a cosmic dance that would continue as long as curiosity drove the human spirit.

The terrestrial twin continued to reside in Dante, part of a man who had touched the stars. The wandering twin, forever immortalized in scientific data, now had a new role as a cornerstone in Dante's new Arizona Astrobiology Center. Each twin carried the memories of their stellar origins as they propelled humanity into a future where the sky was not a limit but an invitation.

In that infinitesimal but immeasurable moment, Dante sensed the unity of all things—the cosmic and the intimate, the ancient and the immediate, the minuscule and the infinite. United at last, the carbon twins settled into their newfound roles, forever entwined in the grand tapestry of life and discovery.

ACKNOWLEDGMENTS

This memoir is a tribute to the collective efforts of remarkable individuals who have been part of my life's journey, shaping it in profound ways. I extend my heartfelt thanks to those who played pivotal roles, each contributing uniquely to its creation.

My deepest gratitude extends to my wife, Kate, and sons, Xander and Griffin, for their unwavering support and understanding throughout this writing journey. You are my rock and inspiration, and I am proud to dedicate this book to you.

My mom, Paula, and my stepfather, Paul, were a constant source of boundless encouragement. During my long business trips, you both stepped up to take care of the family with love and care. Your belief in me has been a source of strength, making you my biggest fans in every endeavor. Your love has been the foundation of my success, and I am forever grateful for the immense role you've played in shaping my life.

My brothers, Nick and Matt, are an inseparable part of my life's journey. Your love, encouragement, and camaraderie have been invaluable to me. Thank you for always being there, for cheering me on, and for being the best brothers anyone could ask for.

My in-laws, Dean and Joann Crombie, welcomed me into the family with warmth and love. Your encouragement meant the world to me. Dean's keen interest in OSIRIS-REx, as an aerospace engineer, was truly inspiring, and witnessing his dream come true was a remarkable experience. Joann's incredible care for our sons while I was away on countless trips has been a blessing beyond measure.

I want to remember and honor Dr. Michael Drake, my mentor and friend. OSIRIS-REx was his dream, and his passion for exploration guided us all. Sadly, he passed away just four short months after we won the contract to fly the mission. Dr. Drake's legacy lives on in the accomplishments of the OSIRIS-REx mission, and I am grateful for the impact he had on my life.

I am indebted to Ashley Stimpson, my exceptional collaborator and writing mentor, whose guidance and expertise have been transformative. With her support, I successfully transitioned from an academic writer to crafting stories for a broader audience. Ashley's exceptional project management skills ensured that the memoir stayed on track, and her insightful feedback helped refine the narrative to resonate with readers.

The peer reviewers of this memoir played an instrumental role in shaping its final form. Their thoughtful reviews and constructive feedback provided invaluable insights that helped refine the narrative and ensure its accuracy and authenticity. Catherine (Cat) Wolner, Arlin Bartels, and Robert (Bob)

Peterson brought diverse perspectives and expertise, each contributing unique viewpoints to the manuscript. Their attention to detail elevated the memoir to new heights.

To Lauren Sharp, my agent, and the entire team at Aevitas Creative Management, I extend my heartfelt gratitude for your unwavering belief in this memoir and for your dedicated efforts in ensuring its journey to fulfillment.

To Maddie Caldwell, my editor, and the wonderful team at Grand Central Publishing, I am profoundly thankful for the invaluable guidance provided throughout the development of this manuscript. Your thoughtful input has been instrumental in shaping the memoir, and your infectious enthusiasm has motivated me to craft it into the very best version possible.

To the main characters of this memoir, Heather Enos, Ed Beshore, Dani DellaGiustina, Arlin Bartels, Estelle Church, and Carl Hergenrother, thank you for being the driving force behind the OSIRIS-REx mission. Your passion for exploration, dedication, and expertise have enriched my writing with the wonders of space. In this memoir, I developed your characters to personify many elements of the team, representing the essential partners of the University of Arizona, Lockheed Martin, and the NASA Goddard Space Flight Center.

To my key advisors on the OSIRIS-REx mission team, including Carina Bennett, Rich Burns, Harold Connolly, Mike Donnelly, Jason Dworkin, Dave Everett, Sandy Freund, Ron Mink, Mike Moreau, Mike Nolan, and Anjani Polit, thank you for your tireless efforts and collaboration. Your combined expertise made the mission a resounding success.

I am grateful to the entire OSIRIS-REx team, past and present, for making the encounter with Bennu possible. Our roster is too large to include here, but I especially acknowledge the science contributions from Coralie Adam, Sara Balram-Knutson, Olivier Barnouin, Kris Becker, Tammy Becker, Beau Bierhaus, Olivia Billett, Brent Bos, Bill Boynton, Keara Burke, Steve Chesley, Phil Christensen, Ben Clark, Beth Clark, Mike Daly, Christian Drouet d'Aubigny, Josh Emery, Chuck Fellows, Mark Fisher, Dathon Golish, Vicky Hamilton, Karl Harshman, Erica Jawin, Hannah Kaplan, Jay McMahon, Dennis Reuter, Bashar Rizk, Ben Rozitis, Andy Ryan, Dan Scheeres, Jeff Seabrook, Amy Simon, and Kevin Walsh.

Professor Carl DeVito, my undergraduate advisor, had a profound impact on my life and career. It was during my work on SETI (Search for Extraterrestrial Intelligence) under your guidance that my eyes were opened to the wonders of the stars and the vast possibilities of space exploration.

I want to extend my deep appreciation to Dr. Carleton Moore for the incredible opportunity he provided me with. His trust in my abilities allowed me access to the vault at ASU, an experience that was both awe-inspiring and humbling.

Special thanks to my Antarctic expedition teammates, tentmate Danny Glavin, team leader Nancy Chabot, mountaineers John Schutt and Jamie Pierce, and other members of ANSMET 2002. Your camaraderie, support, and resilience made the journey to the frozen continent unforgettable.

To the readers of this memoir, thank you for joining me on this journey.

REFERENCES

PROLOGUE

Farnocchia, Davide et al. "Ephemeris and hazard assessment for near-Earth asteroid (101955) Bennu based on OSIRIS-REx data." *Icarus* 369 (2021): 114594.

CHAPTER 1

Albee, Arden L.; Ray E. Arvidson; and Frank D. Palluconi. "Mars Observer mission." *Journal of Geophysical Research: Planets* 97, no. E5 (1992): 7665–7680.

Cocconi, Giuseppe and Philip Morrison. "Searching for interstellar communications." *Nature* 184 (1959): 844–846.

DeVito, Carl L. and Richard T. Oehrle. "A language based on the fundamental facts of science." *Journal of the British Interplanetary Society* 43, no. 12 (1990): 561–568.

Dick, Steven J. "The search for extraterrestrial intelligence and the NASA High Resolution Microwave Survey (HRMS): Historical perspectives." *Space Science Reviews* 64, nos. 1–2 (1993): 93–139.

Drake, Frank. "Project Ozma." *Physics Today* 14, no. 4 (1961): 40.

Drake, Frank. "The search for extraterrestrial intelligence." *Omni* 15, no. 1 (1992): 6–7.

Klein, Harold P. et al. "The Viking Mission search for life on Mars." *Nature* 262, no. 5563 (1976): 24–27.

Kuiper, Gerard P. "Titan: A Satellite with an Atmosphere." *Astrophysical Journal* 100 (1944): 378.

Kuiper, Gerard P. "Infrared Spectra of Planets." *Astrophysical Journal* 106 (1947): 251.

Kuiper, Gerard P. "The second satellite of Neptune." *Publications of the Astronomical Society of the Pacific* 61, no. 361 (1949): 175.

Kuiper, Gerard P. "The fifth satellite of Uranus." *Publications of the Astronomical Society of the Pacific* 61, no. 360 (1949): 129.

Kuiper, Gerard P. *Photographic Lunar Atlas.* Chicago: University of Chicago Press, 1960.

Raulin-Cerceau, Florence. "The pioneers of interplanetary communication: From Gauss to Tesla." *Acta Astronautica* 67, nos. 11–12 (2010): 1391–1398.

Soffen, Gerald A. and A. Thomas Young. "The Viking missions to Mars." *Icarus* 16, no. 1 (1972): 1–16.

CHAPTER 2

Boynton, W. V. et al. "Science applications of the Mars Observer gamma ray spectrometer." *Journal of Geophysical Research: Planets* 97, no. E5 (1992): 7681–7698.

Christensen, Philip R. et al. "Thermal emission spectrometer experiment: Mars Observer mission." *Journal of Geophysical Research: Planets* 97, no. E5 (1992): 7719–7734.

Cunningham, Glenn E. "The Tragedy of Mars Observer." *IFAC Proceedings Volumes* 29, no. 1 (1996): 7498–7503.

Lauretta, Dante S.; Daniel T. Kremser; and Bruce Fegley Jr. "The rate of iron sulfide formation in the solar nebula." *Icarus* 122, no. 2 (1996): 288–315.

Lawler, Andrew. "Astrobiology Institute picks partners." *Science* 280 (1998): 1338.

Levy, David H.; Eugene M. Shoemaker; and Carolyn S. Shoemaker. "Comet Shoemaker–Levy 9 Meets Jupiter." *Scientific American* 273.2 (1995): 84–91.

Mayor, Michel and Didier Queloz. "A Jupiter-mass companion to a solar-type star." *Nature* 378, no. 6555 (1995): 355–359.

McKay, David S. et al. "Search for past life on Mars: possible relic biogenic activity in Martian meteorite ALH84001." *Science* 273, no. 5277 (1996): 924–930.

Milani, A. et al. "Dynamics of planet-crossing asteroids: Classes of orbital behavior: Project SPACEGUARD." *Icarus* 78, no. 2 (1989): 212269.

"TIME Magazine Cover: Comet Hits Jupiter." *Time*, May 23, 1994. https://content.time.com/time/covers/0,16641,19940523,00.html.

Traub, Alex. "Carolyn Shoemaker, Hunter of Comets and Asteroids, Dies at 92," *New York Times*, September 1, 2021.

Zahnle, Kevin and Mordecai-Mark Mac Low. "The collision of Jupiter and comet Shoemaker-Levy 9." *Icarus* 108, no. 1 (1994): 1–17.

CHAPTER 3

Cassidy, W. A. "Masursky Lecture: Retrospective on the US Antarctic Meteorite Program, Or: FUN and Games with Antarctic Meteorites, Or: Frozen Toes and Frozen Meteorites." *28th Annual Lunar and Planetary Science Conference* 28 (1997): 213.

Drake, Michael J. "Presidential address: Presented 2000 August 28, Chicago, Illinois, USA the eucrite/Vesta story." *Meteoritics & Planetary Science* 36, no. 4 (2001): 501–513.

Eppler, Dean B. "Analysis of Antarctic logistics and operations data: Results from the Antarctic Search for Meteorites (ANSMET), austral summer season, 2002–2003, with implications for planetary surface operations." *Geological Society of America Special Papers* 483 (2011): 75–84.

Glavin, Daniel P. and Elizabeth Jarrell. *Frozen in Time: Hunting Meteorites in Antarctica for Signs of Life*. London: Olympia Publishers, 2019.

Marvin, Ursula B. "The origin and early history of the US Antarctic search for meteorites program (ANSMET)." *35 Seasons of US Antarctic Meteorites (1976–2010) A Pictorial Guide to the Collection* (2014): 1–22.

Yoshida, Masaru et al. "Discovery of meteorites near Yamato mountains, East Antarctica." *Antarctic Record* 39 (1971): 62–65.

CHAPTER 4

Brownlee, Don. "The Stardust mission: Analyzing samples from the edge of the solar system." *Annual Review of Earth and Planetary Sciences* 42 (2014): 179–205.

Burnett, D. S. et al. "The Genesis Discovery Mission: Return of Solar Matter to Earth." *Space Science Reviews* 105 (2003).

Kuiper, G. P. "Interpretation of Ranger VII records." *Communications of the Lunar and Planetary Laboratory* 4 (1966): 1–70.

Matson, Dennis L.; Linda J. Spilker; and Jean-Pierre Lebreton. "The Cassini/Huygens mission to the Saturnian system." *Space Science Reviews* 104, nos. 1–4 (2002): 1–58.

McEwen, Alfred S. et al. "Mars reconnaissance orbiter's high resolution imaging science experiment (HiRISE)." *Journal of Geophysical Research: Planets* 112, no. E5 (2007).

Pasek, Matthew A. and Dante S. Lauretta. "Aqueous corrosion of phosphide minerals from iron meteorites: A highly reactive source of prebiotic phosphorus on the surface of the early Earth." *Astrobiology* 5, no. 4 (2005): 515–535.

Saunders, R. S. et al. "2001 Mars Odyssey mission summary." *Space Science Reviews* 110 (2004): 1–36.

Smith, Peter H. et al. "The imager for Mars Pathfinder experiment." *Journal of Geophysical Research: Planets* 102, no. E2 (1997): 4003–4025.

Smith, P. H. et al. "Introduction to special section on the Phoenix mission: Landing site characterization experiments, mission overviews, and expected science." *Journal of Geophysical Research: Planets* 113, no. E3 (2008).

Ticha, J. et al. "1999 RQ36." *Minor Planet Electronic Circulars* (1999).

"Top 100 science stories of 2004." *Discover*, January 2005.

CHAPTER 5

Bolton, S. J. and Juno Science Team. "The Juno mission." *Proceedings of the International Astronomical Union* 6, no. S269 (2010): 92–100.

Boynton, W. V. et al. "The Mars Odyssey gamma-ray spectrometer instrument suite." *Space Science Reviews* 110 (2004): 37–83.

Boynton, W. V. et al. "Evidence for calcium carbonate at the Mars Phoenix landing site." *Science* 325, no. 5936 (2009): 61–64.

Christensen, Philip R. et al. "The OSIRIS-REx thermal emission spectrometer (OTES) instrument." *Space Science Reviews* 214 (2018): 1–39.

Daly, M. G. et al. "The OSIRIS-REx laser altimeter (OLA) investigation and instrument." *Space Science Reviews* 212 (2017): 899–924.

Hecht, Michael H. et al. "Detection of perchlorate and the soluble chemistry of Martian soil at the Phoenix lander site." *Science* 325, no. 5936 (2009): 64–67.

Larson, S. et al. "The Catalina sky survey for NEOs." *Bulletin of the American Astronomical Society* 30 (1998): 1037.

Lauretta, D. S. et al. "The OSIRIS-REx target asteroid (101955) Bennu: Constraints on its physical, geological, and dynamical nature from astronomical observations." *Meteoritics & Planetary Science* 50, no. 4 (2015): 834–849.

Lauretta, D. S. et al. "OSIRIS-REx: sample return from asteroid (101955) Bennu." *Space Science Reviews* 212 (2017): 925–984.

Masterson, R. A. et al. "Regolith X-Ray Imaging Spectrometer (REXIS) aboard the OSIRIS-REx asteroid sample return mission." *Space Science Reviews* 214 (2018): 1–26.

Navarro-González, Rafael et al. "Characterization of organics, microorganisms, desert soils, and Mars-like soils by thermal volatilization coupled to mass spectrometry and their implications for the search for organics on Mars by Phoenix and future space missions." *Astrobiology* 9, no. 8 (2009): 703–715.

Nolan, Michael C. et al. "Shape model and surface properties of the OSIRIS-REx target Asteroid (101955) Bennu from radar and light-curve observations." *Icarus* 226, no. 1 (2013): 629–640.

Reuter, D. C. et al. "The OSIRIS-REx Visible and InfraRed Spectrometer (OVIRS): Spectral maps of the asteroid Bennu." *Space Science Reviews* 214 (2018): 1–22.

Rizk, B. et al. "OCAMS: The OSIRIS-REx camera suite." *Space Science Reviews* 214 (2018): 1–55.

Space Studies Board and National Research Council. *Opening New Frontiers in Space: Choices for the next New Frontiers Announcement of Opportunity.* Washington, DC: National Academies Press, 2008.

Stern, S. Alan. "The New Horizons Pluto Kuiper belt mission: An overview with historical context." *New Horizons: Reconnaissance of the Pluto-Charon System and the Kuiper Belt* (2009): 3–21.

Zuber, Maria T. et al. "Gravity field of the Moon from the Gravity Recovery and Interior Laboratory (GRAIL) mission." *Science* 339, no. 6120 (2013): 668–671.

CHAPTER 6

DellaGiustina, D. N. et al. "Overcoming the challenges associated with image-based mapping of small bodies in preparation for the OSIRIS-REx mission to (101955) Bennu." *Earth and Space Science* 5, no. 12 (2018): 929–949.

Fujiwara, Akira et al. "The rubble-pile asteroid Itokawa as observed by Hayabusa." *Science* 312, no. 5778 (2006): 1330–1334.

Hergenrother, Carl W. et al. "The design reference asteroid for the OSIRIS-REx Mission Target (101955) Bennu." *arXiv preprint arXiv:1409.4704* (2014).

Nakamura, Tomoki et al. "Itokawa dust particles: A direct link between S-type asteroids and ordinary chondrites." *Science* 333, no. 6046 (2011): 1113–1116.

Takimoto, Tomoyuki, director. *Hayabusa: The Long Voyage Home.* Toei Company, 2012.

CHAPTER 7

Berry, Kevin et al. "OSIRIS-REx touch-and-go (TAG) mission design and analysis." *36th Annual AAS Guidance and Control Conference* (2013).

Bierhaus, E. B. et al. "The OSIRIS-REx spacecraft and the touch-and-go sample acquisition mechanism (TAGSAM)." *Space Science Reviews* 214 (2018): 1–46.

Brace, Richard et al. "Report on the loss of the Mars Climate Orbiter Mission." *JPL Special Review Board* (1999).

Chesley, Steven R. et al. "Orbit and bulk density of the OSIRIS-REx target Asteroid (101955) Bennu." *Icarus* 235 (2014): 5–22.

Emery, J. P. et al. "Thermal infrared observations and thermophysical characterization of OSIRIS-REx target asteroid (101955) Bennu." *Icarus* 234 (2014): 17–35.

Jewitt, David and Jing Li. "Activity in Geminid parent (3200) Phaethon." *The Astronomical Journal* 140, no. 5 (2010): 1519.

Ryabova, G. O.; V. A. Avdyushev; and I. P. Williams. "Asteroid (3200) Phaethon and the Geminid meteoroid stream complex." *Monthly Notices of the Royal Astronomical Society* 485, no. 3 (2019): 3378–3385.

Scheeres, Daniel J. et al. "The geophysical environment of Bennu." *Icarus* 276 (2016): 116–140.

Walsh, Kevin J. et al. "Assessing the sampleability of Bennu's surface for the OSIRIS-REx asteroid sample return mission." *Space Science Reviews* 218, no. 4 (2022): 20.

Williams, Bobby et al. "OSIRIS-REx flight dynamics and navigation design." *Space Science Reviews* 214 (2018): 1–43.

Yano, Hajime et al. "Touchdown of the Hayabusa spacecraft at the Muses Sea on Itokawa." *Science* 312, no. 5778 (2006): 1350–1353.

CHAPTER 8

Biele, Jens et al. "The landing (s) of Philae and inferences about comet surface mechanical properties." *Science* 349, no. 6247 (2015): aaa9816.

De León, J. et al. "Visible and near-infrared observations of asteroid 2012 DA14 during its closest approach of February 15, 2013." *Astronomy & Astrophysics* 555 (2013): L2.

Emel'yanenko, V. V. et al. "Astronomical and physical aspects of the Chelyabinsk event (February 15, 2013)." *Solar System Research* 47 (2013): 240–254.

Glassmeier, Karl-Heinz et al. "The Rosetta mission: Flying towards the origin of the solar system." *Space Science Reviews* 128 (2007): 1–21.

Kerr, Richard A. "Flipped switch sealed the fate of Genesis spacecraft." *Science* 306, no. 5696 (2004): 587.

Olynick, David; Y-K. Chen; and Michael E. Tauber. "Aerothermodynamics of the Stardust sample return capsule." *Journal of Spacecraft and Rockets* 36, no. 3 (1999): 442–462.

Strange, Nathan et al. "Overview of mission design for NASA asteroid redirect robotic mission concept." *33rd International Electric Propulsion Conference* (2013).

Thomas, Nicolas et al. "The morphological diversity of comet 67P/ Churyumov-Gerasimenko." *Science* 347, no. 6220 (2015): aaa0440.

CHAPTER 9

Lauretta, Dante S. et al. "OSIRIS-REx at Bennu: Overcoming challenges to collect a sample of the early Solar System." In *Sample Return Missions.* Edited by Andrea Longobardo. Amsterdam: Elsevier (2021), 163–194.

Thomas, Daniel; Gerald Schumann; and Marc Timm. "Human Spaceflight Mishap Investigations: Enabling a Better Model for Future NASA and Commercial Investigations." *New Space* 6, no. 4 (2018): 299–307.

CHAPTER 10

Bos, B. J. et al. "In-flight calibration and performance of the OSIRIS-REx touch and go camera system (TAGCAMS)." *Space Science Reviews* 216 (2020): 1–52.

DellaGiustina, D. N. et al. "Properties of rubble-pile asteroid (101955) Bennu from OSIRIS-REx imaging and thermal analysis." *Nature Astronomy* 3, no. 4 (2019): 341–351.

Golish, D. R. et al. "Ground and in-flight calibration of the OSIRIS-REx camera suite." *Space Science Reviews* 216 (2020): 1–31.

Hamilton, V. E. et al. "Evidence for widespread hydrated minerals on asteroid (101955) Bennu." *Nature Astronomy* 3, no. 4 (2019): 332–340.

Hergenrother, C. W. et al. "The operational environment and rotational acceleration of asteroid (101955) Bennu from OSIRIS-REx observations." *Nature Communications* 10, no. 1 (2019): 1291.

Lauretta, D. S. et al. "The unexpected surface of asteroid (101955) Bennu." *Nature* 568.7750 (2019): 55–60.

Lauretta, D. S. et al. "Episodes of particle ejection from the surface of the active asteroid (101955) Bennu." *Science* 366, no. 6470 (2019): eaay3544.

Leonard, Jason M. et al. "OSIRIS-REx orbit determination performance during the navigation campaign." *2019 AAS/AIAA Astrodynamics Specialist Conference* (2020): 3031–3050.

Simon, A. A. et al. "OSIRIS-REx visible and near-infrared observations of the Moon." *Geophysical Research Letters* 46, no. 12 (2019): 6322–6326.

Yamaguchi, Tomohiro et al. "Hayabusa2-Ryugu proximity operation planning and landing site selection." *Acta Astronautica* 151 (2018): 217–227.

CHAPTER 11

Bennett, C. A. et al. "A high-resolution global basemap of (101955) Bennu." *Icarus* 357 (2021): 113690.

Berry, Kevin et al. "Revisiting OSIRIS-REx touch-and-go (TAG) performance given the realities of asteroid Bennu." *Annual AAS Guidance, Navigation and Control Conference* (2020).

Burke, Keara N. et al. "Particle size-frequency distributions of the OSIRIS-REx candidate sample sites on asteroid (101955) Bennu." *Remote Sensing* 13, no. 7 (2021): 1315.

Enos, H. L. and D. S. Lauretta. "A rendezvous with asteroid Bennu." *Nature Astronomy* 3, no. 4 (2019): 363–363.

Morota, T. et al. "Sample collection from asteroid (162173) Ryugu by Hayabusa2: Implications for surface evolution." *Science* 368, no. 6491 (2020): 654–659.

CHAPTER 12

Daly, M. G. et al. "Hemispherical differences in the shape and topography of asteroid (101955) Bennu." *Science Advances* 6, no. 41 (2020): eabd3649.

DellaGiustina, D. N. et al. "Variations in color and reflectance on the surface of asteroid (101955) Bennu." *Science* 370, no. 6517 (2020): eabc3660.

Kaplan, H. H. et al. "Bright carbonate veins on asteroid (101955) Bennu: Implications for aqueous alteration history." *Science* 370, no. 6517 (2020): eabc3557.

Mario, C. E. et al. "Ground Testing of Digital Terrain Models to Prepare for OSIRIS-REx Autonomous Vision Navigation Using Natural Feature Tracking." *The Planetary Science Journal* 3, no. 5 (2022): 104.

Rozitis, B. et al. "Asteroid (101955) Bennu's weak boulders and thermally anomalous equator." *Science Advances* 6, no. 41 (2020): eabc3699.

Rozitis, B. et al. "High-Resolution Thermophysical Analysis of the OSIRIS-REx Sample Site and Three Other Regions of Interest on Bennu." *Journal of Geophysical Research: Planets* 127 no. 6 (2022): e2021JE007153.

Scheeres, D. J. et al. "Heterogeneous mass distribution of the rubble-pile asteroid (101955) Bennu." *Science Advances* 6, no. 41 (2020): eabc3350.

Simon, Amy A. et al. "Widespread carbon-bearing materials on near-Earth asteroid (101955) Bennu." *Science* 370, no. 6517 (2020): eabc3522.

CHAPTER 13

Barnouin, O. S. et al. "Geologic context of the OSIRIS-REx sample site from high-resolution topography and imaging." *The Planetary Science Journal* 3, no. 4 (2022): 75.

Berry, Kevin et al. "Contact with Bennu! Flight performance versus prediction of OSIRIS-REx 'TAG' sample collection." *AIAA SCITECH 2022 Forum* (2022).

Levine, Andrew H.; Daniel Wibben; and Samantha M. Rieger. "Trajectory Design and Maneuver Performance of the OSIRIS-REx Low-Altitude Reconnaissance of Bennu." *AIAA SCITECH 2022 Forum* (2022).

Norman, C. D. et al. "Autonomous Navigation Performance Using Natural Feature Tracking during the OSIRIS-REx Touch-and-Go Sample Collection Event." *The Planetary Science Journal* 3, no. 5 (2022): 101.

CHAPTER 14

Lauretta, D. S. et al. "Spacecraft sample collection and subsurface excavation of asteroid (101955) Bennu." *Science* 377, no. 6603 (2022): 285–291.

Walsh, Kevin J. et al. "Near-zero cohesion and loose packing of Bennu's near subsurface revealed by spacecraft contact." *Science Advances* 8, no. 27 (2022): eabm6229.

INDEX

ABOUT THE AUTHOR

Dante Lauretta is a Regents Professor of Planetary Science and Cosmochemistry in the Lunar and Planetary Laboratory at the University of Arizona. He is also the director of the Arizona Astrobiology Center. He holds dual bachelor's degrees in science and humanities from the University of Arizona and a PhD in Earth and Planetary Sciences from Washington University in St. Louis, where he developed a deep fascination with understanding the mysteries of the cosmos.

Dante's leadership has been instrumental in steering ambitious space missions. As the principal investigator for NASA's OSIRIS-REx mission to the asteroid Bennu, he played a pivotal role in a historic moment for humanity by collecting a sample from the asteroid's surface, offering unprecedented insights into our solar system's history. He is an expert in the analysis of extraterrestrial materials, including asteroid samples, meteorites, and comet particles.

Beyond his academic achievements, Dante has a passion for communicating complex scientific concepts to the general public. Moreover, he fosters the advancement of the next generation of scientists, engineers, and other space leaders through mentorship and taught coursework. He actively advocates for STEM education, encouraging young minds to embrace science and space exploration.

During his leisure time, Dante enjoys spending quality moments with his family, engaging in outdoor activities, and finding solace in the natural beauty of our planet.